and Main Machinery in Motorships

by N.E Chell
CEng, FIMarEST

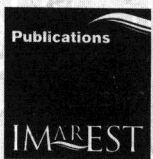

Publications

IMAREST

CW01467589

Published by IMarEST
The Institute of Marine Engineering, Science and Technology
80 Coleman Street • London • EC2R 5BJ

www.imarest.org

A charity registered in England and Wales
Registered Number 212992

First published in 1999 by The Institute of Marine Engineers (now IMarEST)
Reprinted 2004 and 2007
This reprint 2010

ISBN 1-902536-16-9 paperback

British Library Cataloguing-in-Publication Data
A catalogue record for this book is available from the British Library

Contents

Contents

Acknowledgements

The author gratefully acknowledges the following companies for providing material for illustrations.

ABB Turbo Systems Ltd.
Alfa Laval Marine and Power AB
Hamworthy Marine Ltd.
MAN-B&W Diesel
MTU
Standard Piston Ring Co Ltd.
Wärtsilä NSD
Westalia Separator AG

Introduction

Many changes have taken place within the marine industry due to the rapid advances in modern technology, particularly in engine design and shipboard practice. Medium speed engines have become larger and are installed as main propulsion engines in larger vessels, particularly passenger vessels. These engines are highly rated and can burn high viscosity residual fuels.

Slow speed crosshead engines have also changed and the three remaining manufacturers all produce engines of similar design - long stroke, uniflow scavenged with a central exhaust valve. While medium speed builders have added larger engines to their range, slow speed manufacturers have produced smaller bore models. Engine builders have also merged to provide a wider range of engines while remaining competitive.

This new edition provides general guidance for the operation and maintenance of machinery in motorships. Certain aspects are covered in greater detail in other parts of the Marine Engineering Practice series and reference has been made to the relevant titles where appropriate.

While this book is intended to give guidance, at all times the equipment manufacturers operation and maintenance manuals should be referred to. The chapter on electrical machinery sets out guidelines for good practice as electrical maintenance becomes the responsibility of one of the engineers on ships which don't carry electricians.

Changes to shipboard practice have taken place due to anti-pollution regulations and, with more emphasis placed on pollution and environmental protection, it is important that ship's staff are aware of the regulations concerning the disposal of wastes, be it bilge water, sludge, sewage or garbage. The current and proposed annexes to Marpol 73/78 are discussed in chapter 6.

Safety

A ship can be a dangerous place without adequate training and awareness. Along with the many confined spaces, deck plates can be slippery and the vessel may also be rolling or pitching. Extra vigilance and care is essential.

When working in the machinery space or on deck cotton overalls and steel toe-cap shoes or boots should be worn. Nylon overalls should never be worn. Ear protection is also required in machinery spaces and hard hats should be worn during maintenance periods, or out on deck, when people may be working aloft. Goggles should also be worn when chiselling or grinding and dust masks are important if any work is going to create airborne particles. Signs warning of situations where ear and eye protection should be worn must be placed in prominent positions.

Maintenance operations can cause an engineer to come into contact with fuel and lubricants and prolonged exposure should always be avoided. Barrier creams provide some protection but gloves should also be worn. Maintenance equipment should be kept in good condition, a worn spanner can cause a set of grazed knuckles at the least. Always ensure plugs and cables on electrical equipment are in good condition and, if there is any doubt about a piece of equipment do not use it.

When working on an item of machinery ensure that it cannot be started up. Inform others that it is being worked on. For a diesel engine ensure the air start valve is locked shut and a prominent notice put on the start handle. If battery started, disconnect the battery. For electrical machinery isolate the machine at the switchboard and remove the fuses. If possible lock the isolator in the 'off' position and place a prominent notice stating that work is in progress. Before starting work, double check the machine cannot be started. For work on electrical systems above 1kV a 'permit to work' should be used.

Certain operations on board ship are hazardous, such as burning and welding, entering enclosed spaces, working aloft and working on electrical equipment. In order to identify the hazards and eliminate or minimise the risks they pose, a 'permit to work' system should be used. Care should always be taken when entering enclosed spaces. On many ships the emergency fire pump may be in a special compartment that can be closed. In such cases it is usual to have an air supply piped into the space, which should be run for several minutes before entering the space. With enclosed space, such as ballast and bunker tanks, a 'permit to work' should be used.

The 'permit to work' documents the tasks that are to be carried out before entering the space, such as ventilating the space and testing the atmosphere. Once the tasks have been completed the permit is signed by the person in charge. The permit should list the person in charge and who will carry out the work, as well as a period of validity. Samples of the atmosphere should be tested at varying depths and from as many opening as possible. Ventilation should be stopped prior to testing the atmosphere in order to get a representative sample.

After a period of maintenance in port all equipment that has been used needs to be securely stowed prior to sailing. A lot of the equipment used for main engine overhauls is large and heavy and if left unsecured can cause damage, not just to other equipment but also to personnel. Guards must always be replaced on machines.

Safety posters should be placed around to remind people of the dangers. The Code of Safe Working Practices for Seamen and Department of Transport notices to Mariners or M notices are a source of safety information and any relevant notices should be circulated to all ship's staff. Copies of these can be obtained from the UK Marine and Coastguard Agency (MCA).

1 Machinery Arrangements

1.1 Choice of Optimum Engine Type

When building a ship the shipowner must first decide which type of engine to install. The type of ship, i.e. ferry, tanker, container etc., determines the hull form and, to some extent, the operating profile. The type of ship and its operating profile are not the only factors that need to be considered when choosing the main propulsion machinery. Some other factors are:

- Reliability
- Availability
- Maintainability
- Installation costs
- Cost of consumables
- Flexibility
- Machinery costs
- Size of machinery space

Installation costs

The type of machinery determines the services and support arrangements. All these will add to the overall installation costs.

Cost of consumables

The fuel and lubricating oil consumption, as well as the availability of these consumables, may influence the selection of a specific engine type. For instance, can you guarantee getting a high quality fuel or specialist lubricant in every port, or must you carry larger quantities? The costs need to be compared on a daily basis.

Availability

Manning, fuels and lubricants, maintenance etc., all increase the daily running costs of a vessel. If these are high then maximum availability is required. Availability is usually dictated by the ship type, for example, a ferry which is operating to a strict timetable.

Flexibility

There are several factors to be considered. How flexible a vessel needs to be depends on the type of vessel itself and its intended operation. For instance, is a controllable pitch (CP) propeller required which would allow versatile, rapid movement - a prerequisite for a ferry but not for a very large crude carrier (VLCC).

Many smaller vessels have a two stroke engine directly coupled to a CP propeller. This eliminates the need for a reduction gearbox and means the

3

engine will not require reversing gear. These non reversing engines often have a power take off (PTO) for electrical generation. At sea the electricity is generated by the main engine, which will usually have a lower specific fuel oil consumption (sfoc) than the generator engines.

Once the type of engine has been decided, the number of engines required must be determined. Two engines, shafts and propellers can provide some redundancy, however, will increase both installation and operating costs. A single propeller would give a more efficient stern.

Machinery costs

The machinery lifetime needs to be considered. For a typical ship life of 20 years and an average engine operation of say, 7000 hours per year, an engine life of 140 000 hours is required. Also associated with the initial machinery costs are the spares that will be required over this period.

Engine Rating

Once the basic ship type and its main particulars, such as speed, tonnage, length, breadth and draught have been set, the power required to achieve the service speed must be determined. The theory of powering is a lengthy subject beyond the scope of this volume, and engine manufacturers have diagrams to determine the installed power and optimum propeller diameter and revolutions. An example is given below.

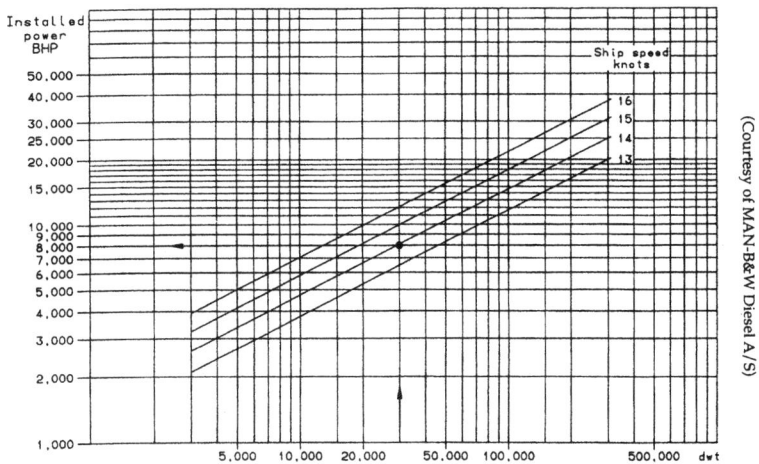

Figure 1. Installed Power Curve for Tanker

(30 000 dwt tanker with a service speed of 14 knots)

4

Using Figure 1, an estimated installed power of 8000bhp is given. Figure 2 is then used to determine the optimum propeller diameter.

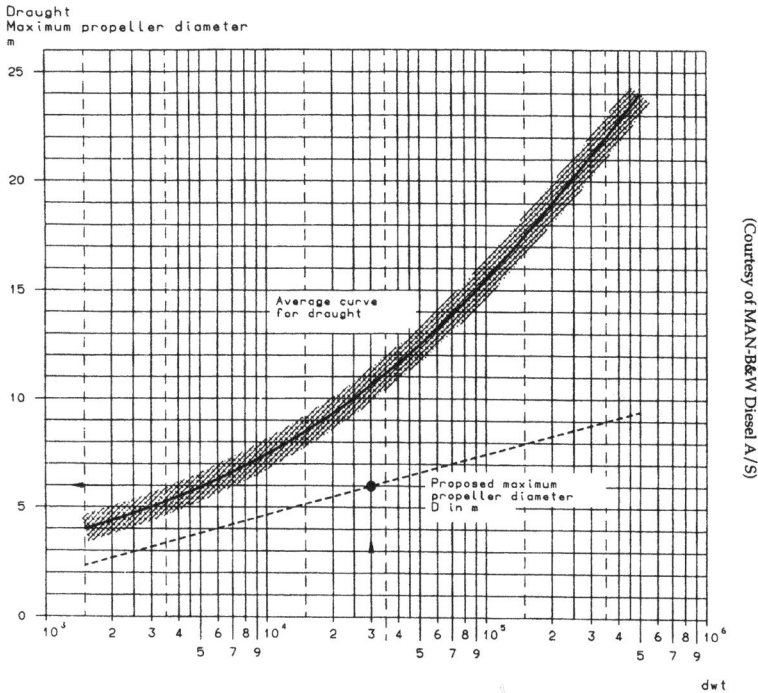

Figure 2. Optimum Propeller Diameter

From Figure 2, it can be seen that a propeller diameter of 6.0 metres is required. Finally, using Figure 3, the optimum engine revolutions can be determined.

This gives a propeller speed of 100 rev/min. For a 30 000 dwt tanker with a service speed of 14 knots, an engine of approximately 8000bhp, driving a 6.0 metre diameter propeller at 100 rev/min is required. This is only an estimate, and engines cover a fairly broad power range, therefore, a manufacturer will probably have more than one engine to satisfy the power requirements. The range of engine outputs for MAN-B&W MC engines are shown in Figure 4. Factors such as maximum permissible propeller diameter and size of engine room also need to be considered.

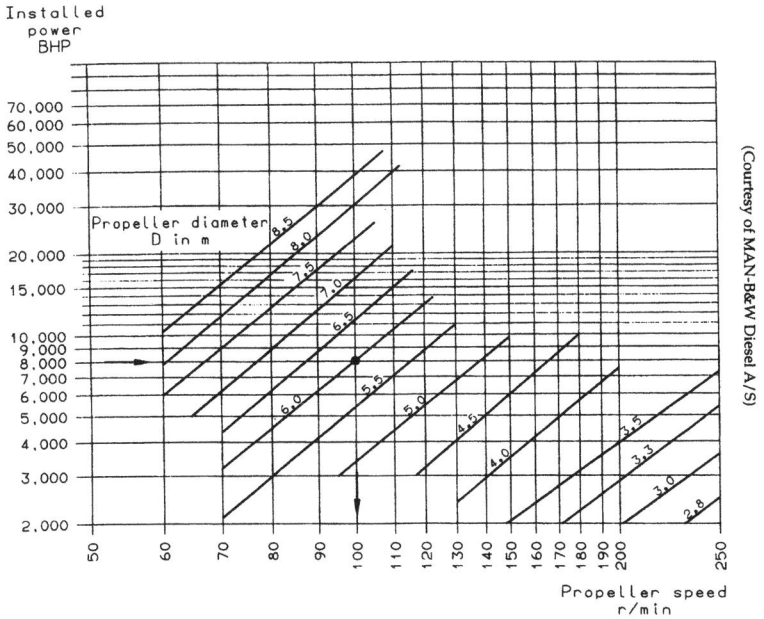

Figure 3. Propeller Speed Curve for Four Bladed Propeller

Engine manufacturers quote a maximum continuous rating for an engine. This is a combination of 100 per cent power, 100 per cent engine speed and 100 per cent mean effective pressure (mep).

Fitting a larger and slower propeller would improve efficiency, however, this may not be possible. There is a requirement for a minimum propeller clearance at the stern to reduce vibration. A large percentage of the propeller may be out of the water on ballast passages, negating any gain in sfoc when loaded.

The Load Diagram

This defines the limit of continuous operation for an installed engine with a given propeller. As with the layout diagram, this is a plot of per cent power, per cent engine speed and per cent mep. In this case the 100 per cent power, 100 per cent engine speed and 100 per cent mep are the specified or selected mcr from the layout diagram.

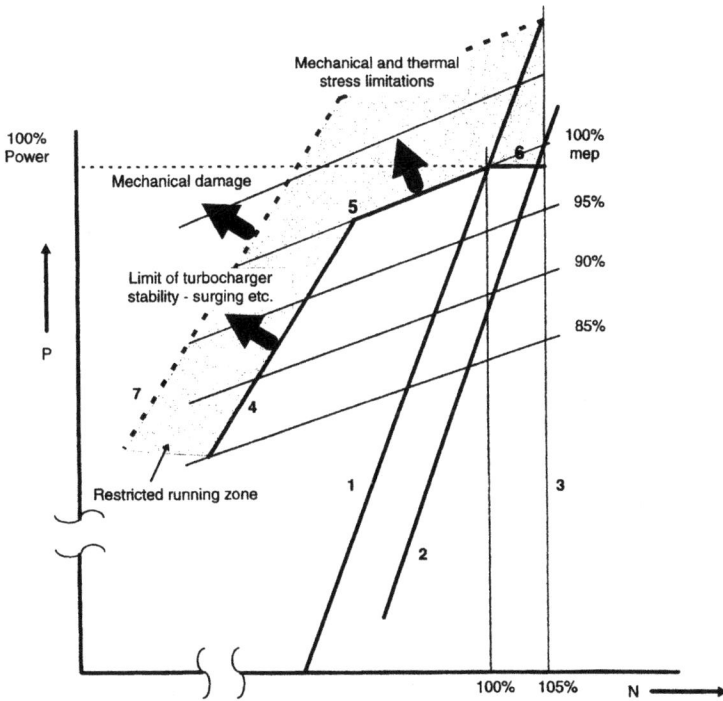

Figure 8. Load Diagram

Line 1.   This represents the engine load on the test bed to simulate the power absorbed by the propeller. This is often referred to as the heavy running curve, with a fouled hull and in heavy weather.

Line 2. This represents the propeller line for a fully loaded new ship, in calm weather and with a clean hull. This curve is sometimes referred to as the light running curve. The line shown is for a fixed pitch (FP) propeller. For a controllable pitch (CP) propeller there is usually a load up curve up to about 50 per cent power. After this point the curve follows that of an FP pitch propeller.

Line 3. This represents the maximum limit of engine speed (rev/min). This is determined by inertia forces and piston speed.

Line 4. This represents the upper limit at any given engine speed at which there will be sufficient air for combustion. Above this limit there is a risk of excessive smoke and turbocharger instability, which would result in surging (refer to turbocharging in Chapter 2).

Line 5. This represents 100 per cent constant mep. Parallel lines are constant mep.

Line 6. This represents the maximum power limit for continuous operation.

Line 7. The shaded area represents an overload operation zone where the engine may only be run for restricted periods.

1.3 Propulsion Machinery - Two and Four Stroke Engines

Internal combustion engines, or heat engines, had been tested for marine use as early as 1888. Dr Rudolf Diesel is considered to be the father of the diesel engine and his 1892 paper entitled 'The theory and construction of a rational heat motor to replace the steam engine and present day heat motors' discussed his new engine cycle. Sulzer Brother's first marine diesel engine was a single cylinder four stroke engine developing 20hp at 160 rev/min, which was first tested in 1898. By today's standards this would be considered a slow speed engine, even though medium speed engines are all four stroke engines. The first two stroke marine engine was introduced in 1905. In general the following can be used as a guide to define the speed of an engine.

* Slow speed: < 200 rev/min
* Medium speed: 200-1000 rev/min
* High speed: > 1000 rev/min

The competition between two and four stroke engines, or to be more precise slow and medium speed engines, was the driving force behind their development. This has resulted in a continual search for higher outputs, smaller and lighter engines and improved specific fuel oil consumption (sfoc).

At one time slow speed engines were favoured because they could burn the higher viscosity residual fuels which were cheaper. This is no longer the case as medium speed engines can now burn fuels up to 700 centistokes.

Some aspects of machinery selection were discussed earlier, however, there are also some physical constraints that need to be considered.

- *Uncontrolled ship motion.* The engine must be able to operate with relatively large angles of list and trim.

- *Ambient conditions.* Systems must be designed for a sea water temperature of at least 35°C, to allow a margin for cooler fouling and ambient air temperature of 40-45°C.

- *Corrosive elements.* These include sea water, sodium in moist engine intake air and vanadium/sodium in fuel which can cause high temperature corrosion.

- *Noise.* Maximum ISO levels of 75 dbA in control rooms and 90 dbA in continuously manned machinery spaces. Ear protection should be worn above 85 dbA.

- *Vibration.* Aft end vibration induced by propeller blade excitation. Axial and lateral shaft vibrations and shaft whirling can also cause damage to gearing or the sterntube bearing.

There are several basic design differences between the two stroke and four stroke engine. These differences can be summarised as follows:

- The vast majority of two stroke engines are of the crosshead design. The piston is connected to the crankshaft via a piston rod, crosshead bearing and guide, and connecting rod. The linear motion of the piston rod allows the underpiston space and crankcase to be separated by a gland. This allows a separate lubrication system for the cylinders and crankcase.

- The slow speed crosshead engine typically has a fabricated bedplate and semi-built crankshaft. Many of the smaller bore crosshead engines now have cast bedplates and A frames or columns. The smaller size allows these items to be cast as one piece which is beneficial because access for welding is difficult. Chapter 2 contains more information on bedplates.

- Modern two stroke engines are uniflow scavenge with an exhaust valve and constant pressure turbocharging. Modern four stroke engines have four valves per head. Smaller four stroke engines are usually pulse turbocharged, although larger modern engines can use constant pressure turbocharging alone or a combination of both pulse and pressure turbocharging.

- The higher breathing capacity of the four stroke engine allows for higher piston speeds. In a two stroke piston speed is restricted to approximately 8m/s due to pressure losses experienced during scavenging. The speed of the two stroke is also limited due to the inertia forces of the heavier components.

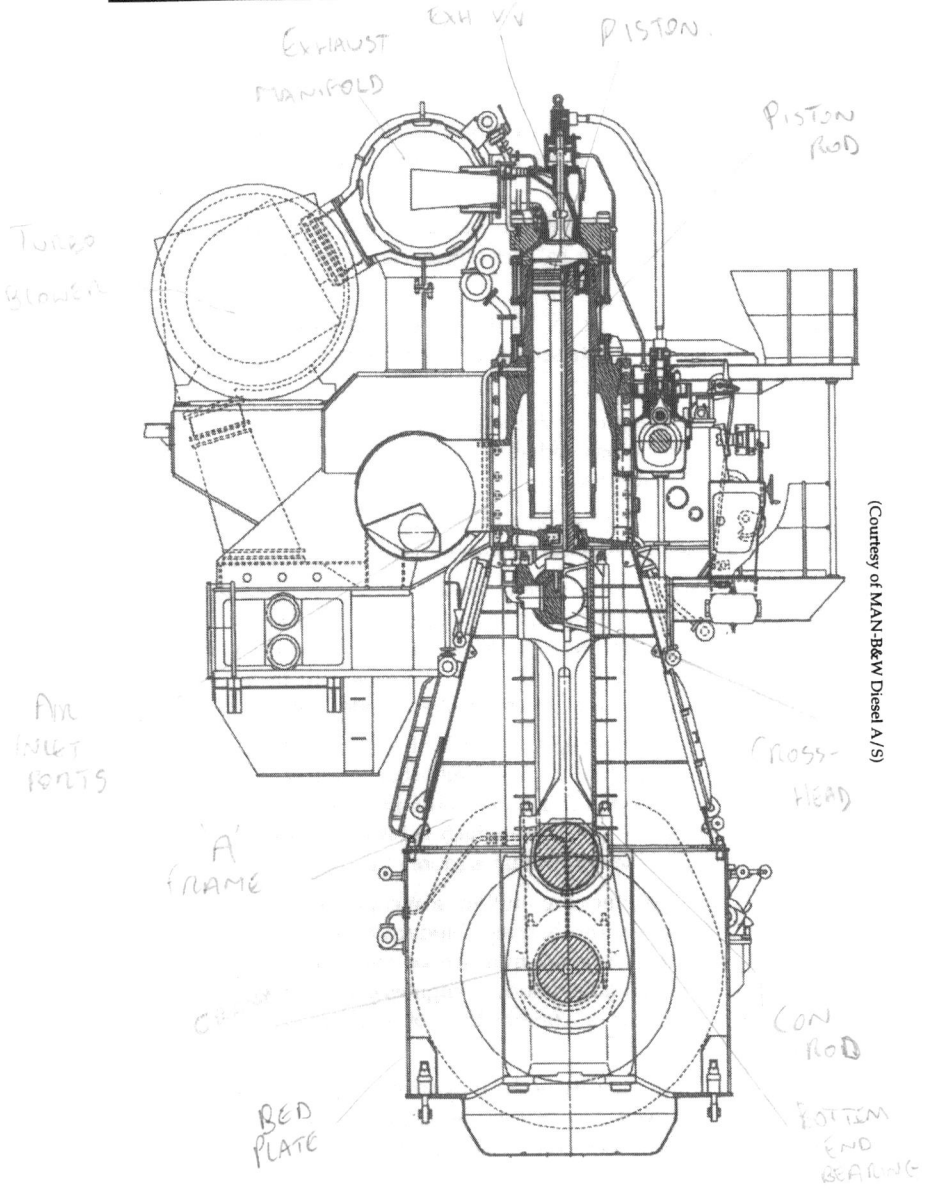

EXHAUST MANIFOLD

EXH V/V

PISTON.

PISTON ROD

TURBO BLOWER

AIR INLET PORTS

'A' FRAME

CROSS-HEAD

CON ROD

BOTTOM END BEARING

BED PLATE

(Courtesy of MAN-B&W Diesel A/S)

Figure 9. MAN-B&W S60-MC Two Stroke Engine Cross Section

14

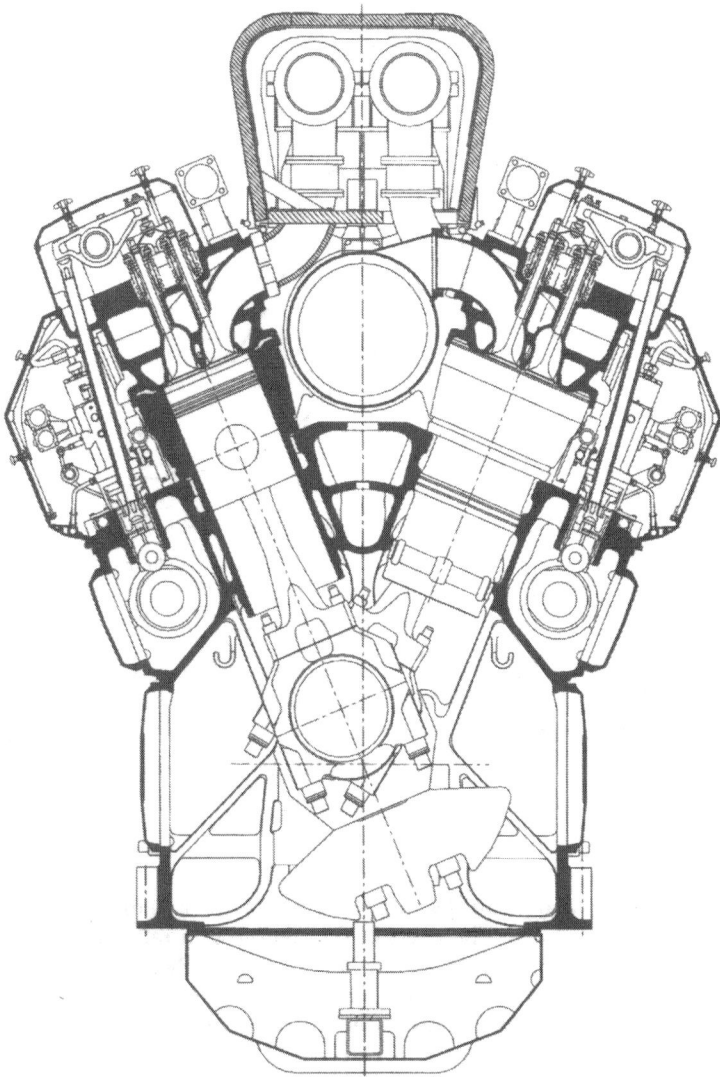

(Courtesy of Wärtsilä NSD)

Figure 10. Wärtsilä NSD W46 Four Stroke Engine Cross Section

<u>The Diesel Cycle</u>

Fuel injection & combustion
occur at constant volume

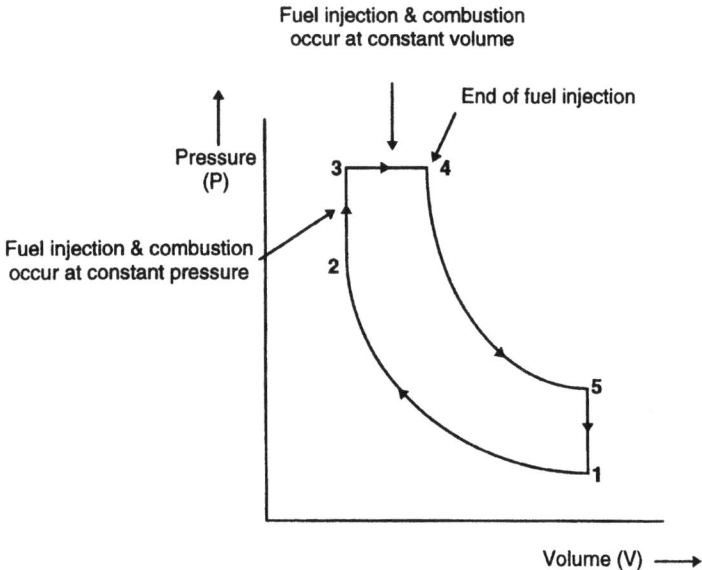

Figure 11. Diesel Cycle

All two and four stroke engines work on the 'dual cycle', which is a combination of the Otto and Diesel cycles. The area of the diagram represents the work done per cycle.

The thermal efficiency of the above cycle depends on three ratios:

i)   Compression ratio $r_v = V_1/V_2$
ii) Pressure ratio $r_p = P_3/P_2$
iii) Cut-off ratio $r_c = V_4/V_3$

The compression ratio is fixed by engine design (i.e. the stroke) and cannot be adjusted by the engineer. The effect of the compression ratio is to produce high pressure in the cylinder at the end of compression. By keeping the rings, liners and valves in good condition the compression pressure will be kept at a maximum. The high pressure results in the high temperature required for efficient combustion.

The pressure ratio is determined by the point of fuel injection. Injection normally occurs about 20° before top dead centre to 20° after top dead centre. If the point of fuel injection is retarded, i.e. occurs later, the peak pressure will be reduced. If the point of fuel injection is advanced then the peak pressure will be increased. High peak pressures should be avoided due to the increased mechanical stresses it puts on the running gear.

The cut-off ratio is determined by the quantity of fuel injected into the cylinder, i.e. by the position of the fuel lever.

Although a high compression ratio and peak pressure will produce a high efficiency, in practice thermal efficiency is also influenced by other factors, such as fuel quality, which may affect the rate of heat release and therefore the efficiency of the turbocharging system.

2 Main Propulsion Machinery – Operation and Maintenance

2.1 General

Although this chapter covers some aspects of the operation and maintenance of main propulsion machinery, it should not replace the manuals issued by the engine manufacturer. These manuals will contain detailed information on the maintenance procedures with exploded drawings, lists of parts that may require renewing, such as seal rings etc., and any special tool requirements.

When preparing to overhaul an engine component the maintenance manual should be studied beforehand. This is particularly important if there are new engineers on board who may not have performed the task before. It is always useful to get all tools ready prior to overhaul to save time.

Some safety aspects have been mentioned in the introduction, but when working on any engine it is vital to ensure it cannot be started during overhaul. Ensure the starting air to the engine has been isolated and that the valves on the air receiver have been locked shut. The drain valve on the air line should be open to ensure the line is vented. All indicator cocks should also be open.

The turning gear should be engaged and in the locked position when working in the crankcase. If need be, the fuses should be removed from the turning gear. There are occasions when someone will be in the crankcase when turning the engine, such as when reassembling a crosshead bearing, and great care should always be taken. The main lubricating oil pumps should be isolated and the fuses removed.

2.2 Running Gear

2.2.1 Cylinder Heads

Cylinder heads, whether on a two or four stroke engine, are complicated castings. They have to house the inlet and exhaust valves, fuel injector, air start valve, relief valve, and indicator cock, as well as incorporating cooling water passages.

Due to the complexity of the casting, care must be taken in the design and manufacture to ensure the fillets have good radii, inspection holes are well compensated and stud holes are bossed. Coolant flow within the head should be as smooth as possible, which is particularly important around high temperature zones such as the injector pocket and exhaust valve cages.

The trend in recent years has been to reduce sfoc. One of the factors involved has been to increase the combustion pressure. This is shown graphically in Figure 12 and results in increased mechanical stress on the cylinder head, as well as higher temperatures.

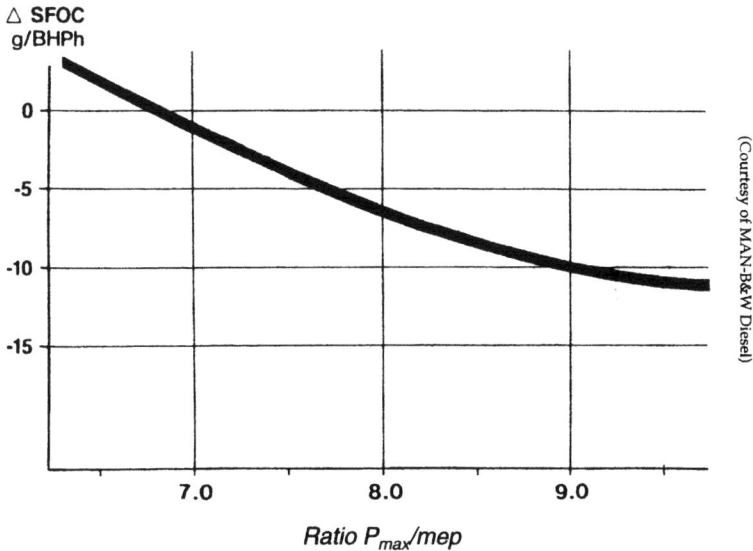

Figure 12. Influence of Ratio P_{max}/mep on sfoc

It is clearly preferable to manufacture the head from cast steel, however, this can result in complications in casting. Cast steel does not flow as freely as cast iron, and may cause porosity. Using cast steel also increases the costs due to the strict quality control required during manufacture. Some heads are forged. These heads have a fine grain structure on their outer surfaces which provides better resistance to cracking. Figures 13 and 14 show cylinder head assemblies for two and four stroke engines respectively.

Both engines have exhaust valves with vanes to rotate the valve and water cooled valve seats. On the four stroke engine the inlet valve is fitted with a mechanical valve rotator (rotocap).

On a two stroke engine the larger size means the head is often made up of at least two parts. This may be a cast steel or forged cover, with a bolted on exhaust valve pocket. Older MAN engines used a two piece head, a cast steel lower section, which was a relatively simple casting, and a cast iron upper part accommodating all the head valves. On large bore engines the larger mass and thicker material reduces the heat transfer rate, leading to a greater temperature gradient and increased thermal stresses.

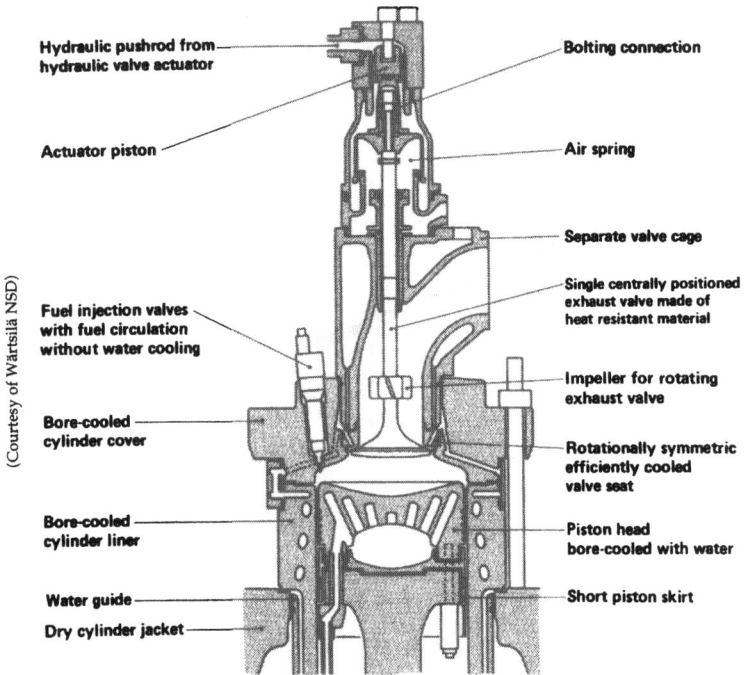

(Courtesy of Wärtsilä NSD)

Hydraulic pushrod from hydraulic valve actuator

Actuator piston

Fuel injection valves with fuel circulation without water cooling

Bore-cooled cylinder cover

Bore-cooled cylinder liner

Water guide

Dry cylinder jacket

Bolting connection

Air spring

Separate valve cage

Single centrally positioned exhaust valve made of heat resistant material

Impeller for rotating exhaust valve

Rotationally symmetric efficiently cooled valve seat

Piston head bore-cooled with water

Short piston skirt

Figure13. Wärtsilä NSD RTA Cylinder Head and Exhaust Valve Assembly

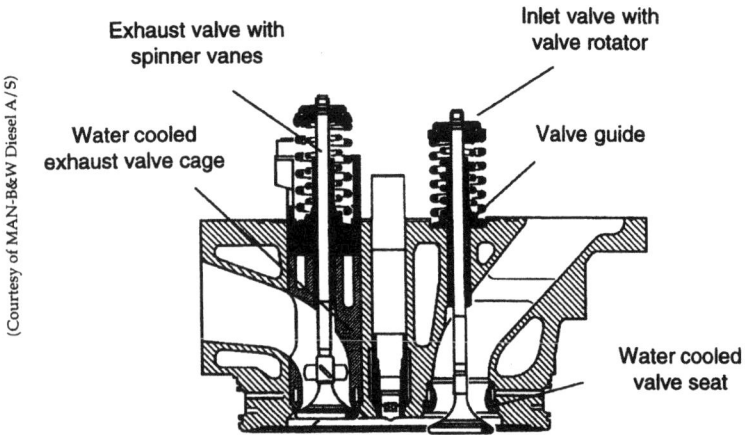

(Courtesy of MAN-B&W Diesel A/S)

Exhaust valve with spinner vanes

Inlet valve with valve rotator

Water cooled exhaust valve cage

Valve guide

Water cooled valve seat

Figure 14. MAN-B&W 48/60 Cylinder Head Assembly

The head is rigidly bolted down, therefore, the high temperature side is in compression. Any cracks will occur on the cooling side and be evident by combustion gas entering the cooling water system. To increase the compressive load the head can accept, large cylinder heads are bore cooled.

As well as thermal stresses, the other main stresses on the cylinder head are shown in Figure 15.

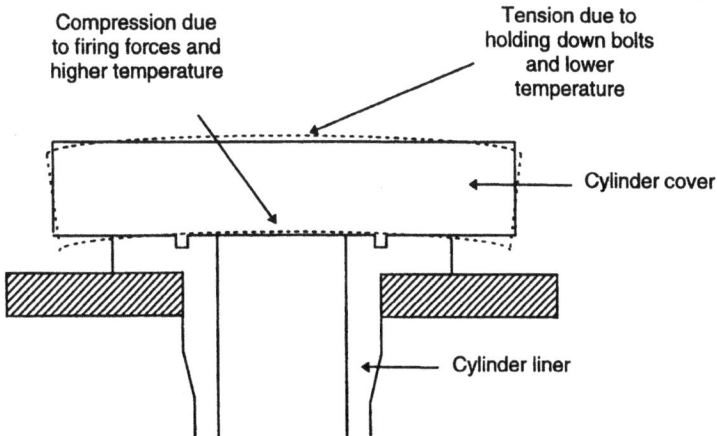

Figure 15. Tensile and Compressive Stresses in a Cylinder Head

The main cylinder head defect is cracking. Possible causes of this are:

Design

i. Inadequate fillets leading to stress concentrations.

ii. Insufficient rigidity and flexing will lead to a higher stress range.

iii. Complex cooling space design allows the formation of air pockets and allows debris to collect, each of which can affect heat transfer and increase thermal stress.

iv. Choice of material and material thickness.

Manufacture

i. Casting defects and locked in stresses.

ii. High surface roughness, which reduces fatigue strength.

Operation

i. The most common cause of failure is through stress corrosion cracking. It is essential to maintain cooling water treatment at the correct level because corrosion reduces fatigue strength.

    ii. Mechanical or thermal overload.

    iii. High peak pressures.

    iv. Thermal shock caused by sudden cooling or load change.

    v. Insufficient cooling. This may happen due to a fault in the cooling system, such as a fouled cooler or defective controller, or could be caused by a build up of dirt and debris in the cooling spaces which reduces the heat transfer.

    vi. Poor combustion - this may lead to point iii (above).

    vii. Over-tightening the head bolts or valve cage bolts.

Correct maintenance and operation are important factors in reducing the risk of cracking in cylinder heads.

It is important that cooling water treatment is kept at the recommended levels. Water tests should be carried out weekly, at least, and more frequently after any water has been drained out during overhaul.

In large bore engines, the engine should be warmed through slowly prior to starting. The cooling water should be heated to almost the normal operating temperature. If the temperature is kept too low, once the engine is running the water will quickly heat up which will increase thermal stresses in the head. If the engine has to be started with the cooling water at low temperatures, the jacket water controller should be operated manually until normal operating temperature is reached.

For slow speed engines the load should be increased gradually. This is not as critical on smaller engines, which have cylinder heads of a smaller mass that warm up fairly uniformly.

Ensure cylinder heads are kept clean internally. At overhaul remove inspection covers for the cooling spaces and flush them out with a high pressure hose to remove debris and deposits.

Bad combustion can lead to increased peak pressures and localised hot spots. It is important, therefore, that fuel injection and treatment systems are kept in efficient condition.

Cylinder heads should be correctly tightened down. For small heads this may be done with a torque wrench allowing the nuts to be tightened down in a certain sequence. On larger heads there may be a special hydraulic 'spider' that tightens all the nuts at once. Whichever method is used, ensure the manufacturer's tightening procedure and torque/pressure setting are adhered to.

Operation with Defective Cylinder Head (Wärtsilä NSD RTA)

A problem may arise where a cylinder needs to be isolated. Wherever possible, the cylinder head should be changed with the spare. If this is not possible then the following actions should be taken, after first stopping the engine:

- Isolate the cooling water to the affected cylinder head.
- Lift the fuel injection pump to cut off the fuel supply to the affected cylinder.
- Lift the actuator pump for the exhaust valve and close off the oil supply. Then vent the air spring.
- Ensure the exhaust valve is left slightly open (this requires a special tool, which is supplied by the manufacturer). This prevents the heat generated during compression from overheating the head.

Once the engine is back in operation care should be taken that other cylinders are not overloaded. Exhaust and cooling water temperatures should be closely monitored. It should be remembered that this is an emergency procedure and the cylinder head should be replaced as soon as possible.

Removal of Cylinder Head (Wärtsilä NSD RTA)

The following applies to Wärtsilä NSD RTA engines but the method is similar for other slow speed engines. The engine manufacturer's maintenance procedures should always be referred to.

- Isolate the cooling water to the affected cylinder head and drain the water from cooling spaces. Make sure the vent is open to allow water to drain.
- Disconnect the following items: the cooling water outlet pipe; the air start pipe and control air pipes from the air start valve; the oil and air pipes for the exhaust valve; and the fuel oil high pressure and circulation pipes.
- Disconnect the expansion piece between the exhaust manifold and valve.
- Clean all cylinder head stud threads and fit hydraulic tensioning device.
- Ensure all the jacks are tightened down and slackened back slightly. This process can be assisted by connecting the hose to the pump.
- Start the pump and gradually increase the pressure to the correct setting. Ensure no connections are leaking.
- Slacken off all cylinder head nuts.
- Remove the tensioning device and all cylinder head nuts. Ensure these are kept in order.
- Remove the cylinder head using the appropriate lifting eyes. On some occasions the head may need some assistance in breaking free from its seat. First check all connections have been removed and no head nuts

have been overlooked. If hydraulic jacks are being used to free the head never use one jack on its own and always use minimal pressure. Excessive force from the jacks may shear the liner locating bolts and dislodge the liner.

Removal of Cylinder Head (Medium Speed Engine)

- Close the cooling water header tank valve and drain the cooling water from the engine.
- Isolate the fuel valve cooling water.
- Turn the engine to top dead centre on firing stroke, so the inlet and exhaust valves are closed.
- Remove the rocker arms and pushrods.
- Disconnect and remove the fuel supply, leakage pipes, cooling water pipes and air start pipe (this arrangement will vary depending upon the engine).
- Disconnect the air inlet and exhaust pipes.
- Remove the cylinder head nuts. This can be done hydraulically or manually.
- Using the appropriate lifting points, remove the cylinder head. Some manufacturers may supply jacking bolts to free the head.

Note: When removing some small pipes ensure the threaded connections are protected to prevent them being damaged during the overhaul period. The same applies to the cylinder stud threads.

Cylinder Head Inspection

- On smaller engines it may be possible to manhandle the head to facilitate cleaning and overhaul. If a stand is not available care should be taken not to damage the head, particularly near the sealing face. The head should be placed on blocks of wood or thick packing.
- Thoroughly clean the combustion space and remove all soot/carbon deposits.
- Using a bright light, carefully inspect in way of valve pockets and openings for any signs of cracks. If in doubt, use a dye penetrant to test for cracks.
- Inspect the combustion chamber for any signs of burning.
- Inspect the valve seats for excessive pitting/shrouding.
- Inspect the head/liner sealing face for any damage.

- Inspect the cooling water space for any sludge build up or corrosion. If necessary flush it out.
- While the head is off replace the air start valve, relief valve and injector(s) with reconditioned spares.
- Replace all joints and seals. Where soft iron sealing rings are fitted these should be replaced. Copper rings can be annealed and reused.

2.2.2   Inlet and Exhaust Valves

Inlet and exhaust valves are subject to severe operating conditions. On a medium speed engine with four valves they make up a large percentage of the combustion chamber surface area. Modern crosshead engines are uniflow scavenged with a large central exhaust valve.

Inlet valves do not generally present any serious problems as they are cooled by the incoming charge air. Exhaust valves are subjected to higher temperatures and their life is limited by deposits and high temperature corrosion from sodium/vanadium compounds (see page 32 for valve problems and maintenance). It is very important to keep the valve seat temperatures below 450°C.

Several factors can influence valve life:

Choice of material

- Strength at high temperature is an important requirement. The valve must have good corrosion and erosion resistance. It is also important that the material is machineable.
- Exhaust valves may be nimonic (80 per cent nickel, 20 per cent chrome), which are often used in highly rated engines, or of heat resistant steel (25 per cent nickel, 12 per cent chrome) with stellite seats (50 per cent cobalt, 30 per cent chrome, 20 per cent tungsten).
- Inlet valves are typically 0.3 per cent chrome and three per cent nickel alloy steel.
- Seat materials vary from manufacturer to manufacturer but are generally an alloyed cast iron, with up to 15 per cent chrome, heat resistant steel or stellite. Stems are usually subjected to a hardening process such as nitriding or chrome or tungsten carbide coated.
- Valve guides are usually pearlitic cast iron that can be surface hardened.

Note: The above gives a guide to some of the materials used. Some manufacturers use their own particular materials.

Cooling

Most of the heat transfer is from the valve to the seat. It is therefore important to have good contact between the seat and valve, particularly at high temperatures. This can be achieved by having a slightly different angle between the seat and the valve to ensure full face contact at operating temperature. Exhaust valves are usually mounted in water cooled cages which, as well as improved cooling, allows for ease of valve maintenance. Valve seats can also be water cooled. Figure 16 shows typical valve seat temperatures.

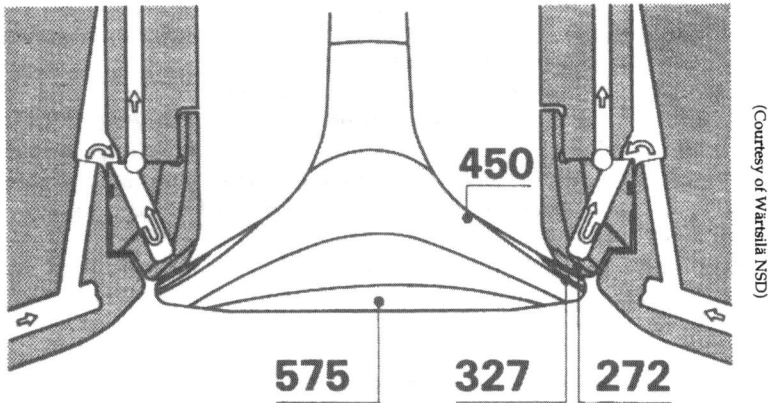

Figure 16. Valve Seat Temperatures (^{o}C) for Wärtsilä NSD RTA58 @ mcr

Only a small amount of heat gets transferred up the stem. This can be increased by using materials with higher thermal conductivity, such as by using sodium filled valves or water cooled stems, though these are not common and are expensive. Excessive stem cooling can also promote corrosion.

Valve rotation

Valve rotation serves two purposes:

i. Helps to dislodge deposits and prevent deposits building up on the seat.

ii. Maintains an even temperature around the valve seat. This increases the service life of the valve by reducing the risk of hot spots developing.

There are basically two methods of rotating the valves; mechanically operated each time the valve opens or closes or by means of the exhaust gas, by fitting a vane or impeller on the valve spindle.

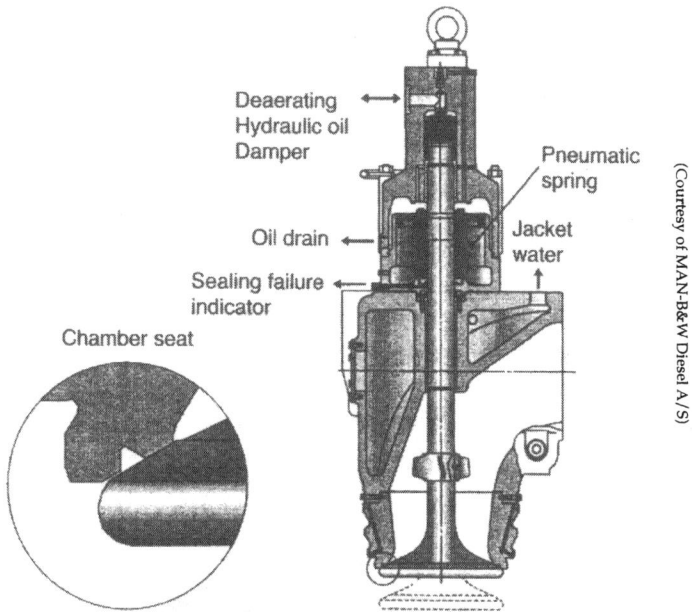

Figure 17. MAN-B&W S-MC-C Exhaust Valve Fitted with Vane

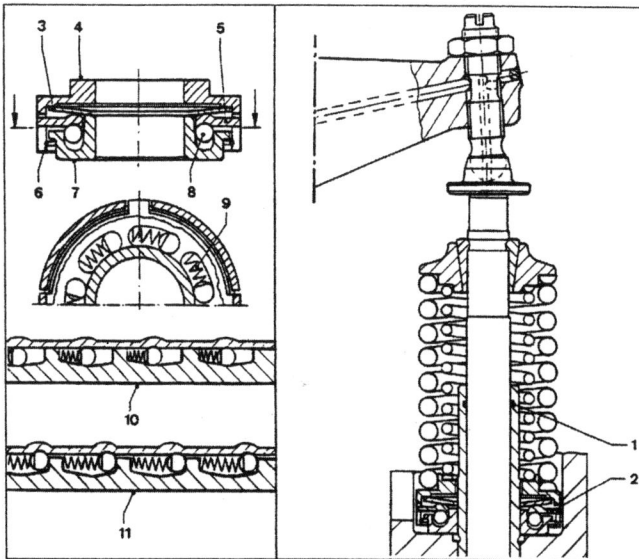

Figure 18. Mechanical Valve Rotator

Operation of valve rotator

In Figure 18, when the valve opens the dished washer (3) is compressed and acts on the steel balls (8). This causes them to move down their inclined grooves which imparts a rotary motion. The cover (4) moves with the dished washer and transmits the rotary motion through the valve spring to the valve. As the valve closes, the force on the dished washer is released and the springs (9) return the balls to their normal position.

Another type of valve rotator is the 'Turnomat'. This rotates the valve on closing which cleans the seat faces in the process.

Figure 19 shows how an even temperature is maintained across the valve seat in a Wärtsilä NSD ZA40 engine.

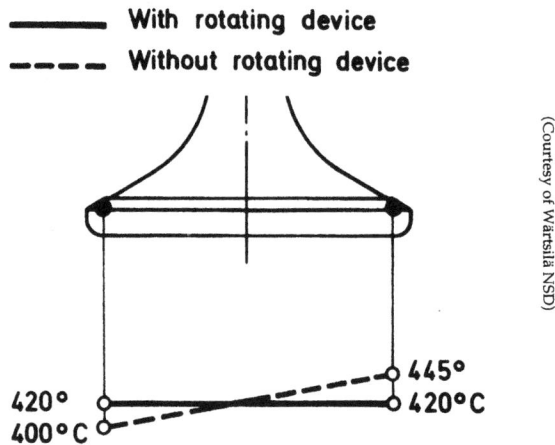

Figure 19. Temperature Differential on Valve Seat

Valve Gear

Valves are traditionally operated by rocker arms and push rods. As the rocker arm operates in an arc the spindle exerts a side thrust on the guide. This results in wear to both the spindle and guide. Increased wear to the guides is usually accompanied by an increase in oil consumption in four stroke engines and, if wear is excessive, gas can be seen blowing up past the guide.

Valve clearance is important as valves will expand as they warm up, particularly exhaust valves. If there is inadequate clearance when the valve expands the valve may not seat properly, if at all. This will lead to increased seat temperatures causing burning of the seating faces.

The valve clearances will change between the hot and cold conditions, therefore, when setting the clearances the manufacturer's figures should be checked to see whether they are for a hot or cold engine.

On large two stroke engines traditional rocker operated valves had particular problems caused by vibration and transverse movement of the springs. To overcome temperature differences in the valve gear temperature compensators had to be fitted.

Modern two stroke uniflow scavenged engines have hydraulically operated valves rather than the traditional rocker arms and push rods method. This allows linear valve motion and eliminates the side thrust and the need for temperature compensators. The elimination of the rocker gear has also led to easier and cheaper maintenance.

Valves are usually closed by springs, although on modern crosshead engines the valve is closed pneumatically using an air spring. The advantage is that vibration is reduced and the need for springs eliminated.

Considering Figure 20, when the exhaust valve is opening from the closed position the air supply is off and the air spring vented. The cam lifts the actuator pump plunger and the pressure acting on the piston causes the exhaust valve to open. To close the valve, compressed air is supplied to the air spring and the control valve opened to vent the oil. This control valve is used to vary the closing of the valve (variable exhaust closing (VEC) control).

Another feature is load dependent pneumatically controlled valve timing which gives variable closing of the inlet valve. This provides a flatter fuel consumption characteristic over the whole load range.

Wärtsilä NSD also utilise electronic VEC on their large crosshead engines. VEC closes the exhaust valve earlier in the cycle which results in an increase in both compression and peak pressures at part load operation. Figures 21 and 22 show how valve lift and cylinder pressures are affected and the reduction in fuel consumption when used with VEC.

VEC is achieved by venting oil from the actuator pump. The valve is then closed by the air spring. The point in the cycle when the valve closes depends on the engine power and is controlled by the charge air pressure and engine speed. VEC operates below 80 per cent mcr and is adjustable over the range 65-80 per cent mcr as shown in Figure 23.

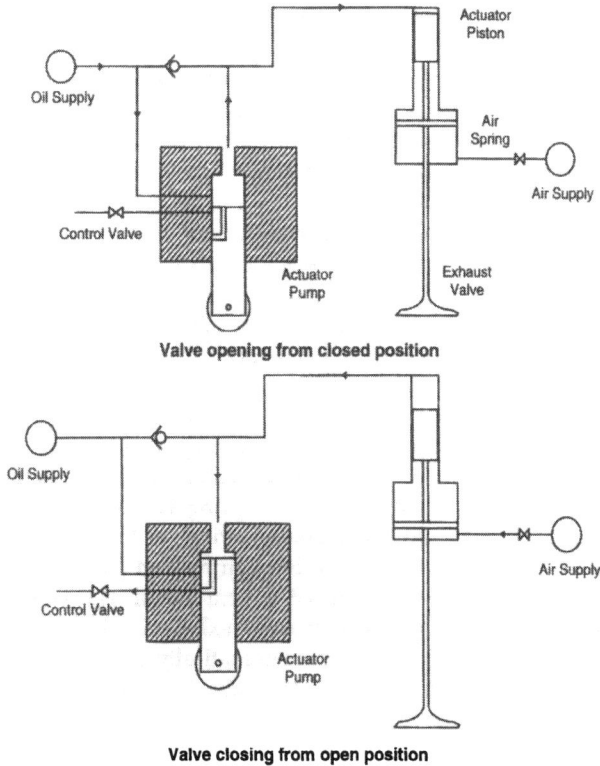

Valve opening from closed position

Valve closing from open position

(Courtesy of Wärtsilä NSD)

Figure 20. Wärtsilä NSD RTA Hydraulically Operated Exhaust Valve

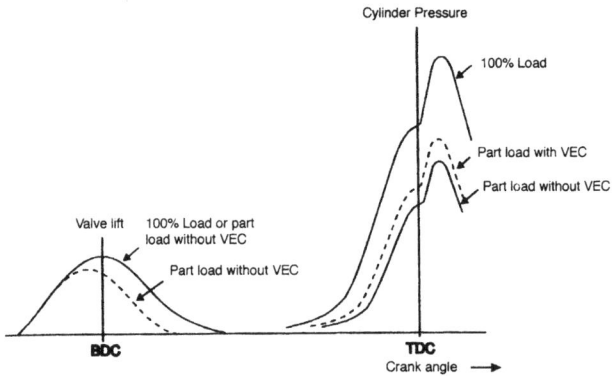

(Courtesy of Wärtsilä NSD)

Figure 21. Valve Lift and Cylinder Pressure

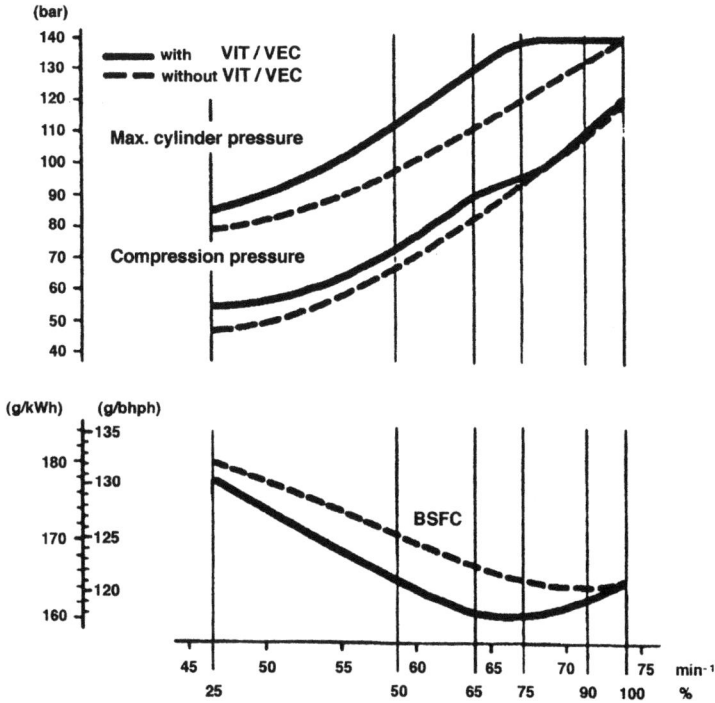

Figure 22. Cylinder Pressures and Fuel Cons

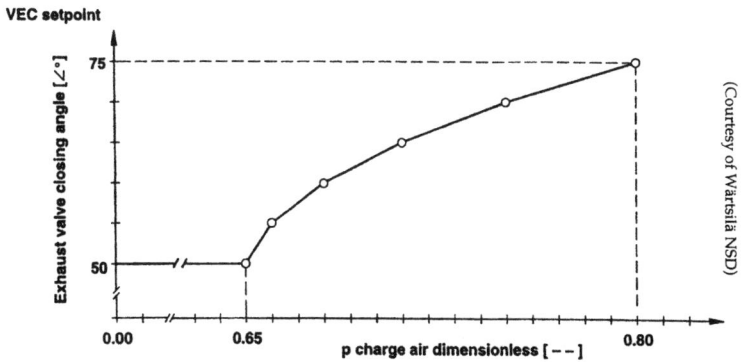

Figure 23. Exhaust Valve Closing Adjustment

(Courtesy of Wärtsilä NSD)

Valve Problems and Maintenance

The main problem valves are subjected to is burning of the sealing face, particularly four stroke engine exhaust valves. This may be due to the following:

- *Valve not closing properly.* This could be due to a build up of deposits on the valve seat or insufficient valve clearance.

- *Poor quality fuel.* Fuel with a high vanadium content may lead to high temperature corrosion (see Figure 24), particularly if sodium is also present. High water content can also lead to valve deposits (see section 2.4 on Fuels and Bunkering).

- *Poor combustion.* This may be due to poor quality fuel or defective injection equipment. Slow burning fuels can cause afterburning when the valve is open.

- *Overheating.* This may be due to overload of the engine resulting in higher exhaust temperatures or restricted coolant flow. Increasing the valve seat temperature increases the risk of high temperature corrosion.

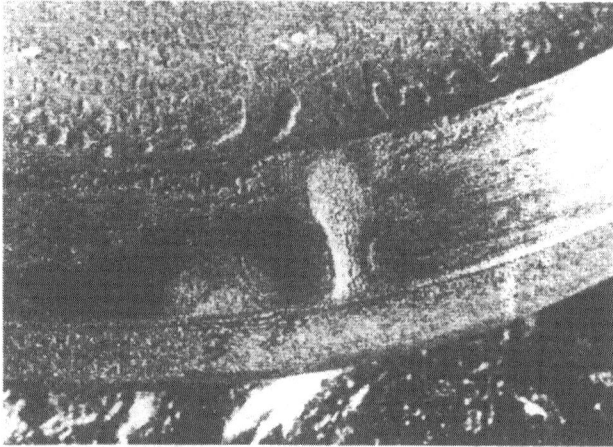

(Courtesy of Wärtsilä NSD)

Figure 24. High Temperature Corrosion on Exhaust Valve Seal Face

Figure 24 shows an exhaust valve with high temperature corrosion on the valve after 2000 hours operation. In severe cases the valve will be completely burned through and rendered unserviceable.

Burning of the valve face results in leakage of gas across the valve seat faces. If small this may not be noticeable but as it gets worse, the exhaust temperature of the affected cylinder will increase. In severe cases the fuel pump will have to be lifted to put the cylinder out of operation.

Leaking valves should be attended to as soon as possible. The high gas velocity leaking across the valve will damage the valve seat.

Shrouding of valve seats can be a problem on smaller engines where the valves have been overhauled several times. The continual grinding process causes a ridge to build up. This can be on both faces or on either the valve or seat, depending which is the hardest. This is shown in Figure 25.

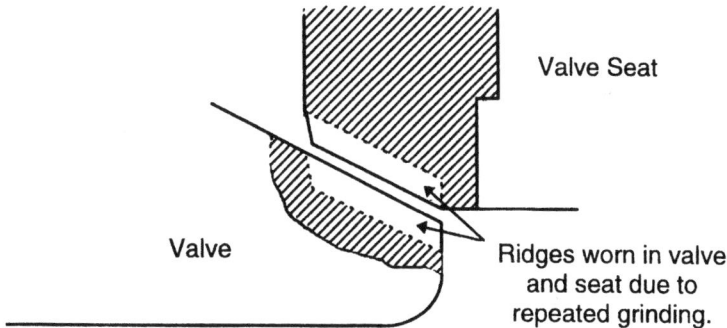

Valve Seat

Valve

Ridges worn in valve and seat due to repeated grinding.

Figure 25. Valve Shrouding

The ridge must be removed by grinding because shrouding reduces the effective opening area of the valve which restricts the gas flow.

Worn valve guides and valve stems can also be a problem. Larger valves may have a direct oil supply to ensure adequate lubrication. Worn guides will increase oil consumption on trunk piston engines and lead to gas blowby which will damage the stem and guide.

MAN-B&W utilise sealing air with oil mist for their MC crosshead engines to minimise wear of the valve stem. This arrangement is shown in Figure 26.

Valve Maintenance

Once the valves have been removed they should be thoroughly cleaned of all carbon and other deposits then visually examined. This also applies to the valve cages or pockets and seats. Valve stem dimensions should be checked against the manufacturer's limits and any worn valves replaced or reconditioned, if possible.

Valve seat faces should be inspected for any ridges, pitting or burning. The extent of any damage will determine whether the valve needs to be sent for reconditioning or can be ground on board.

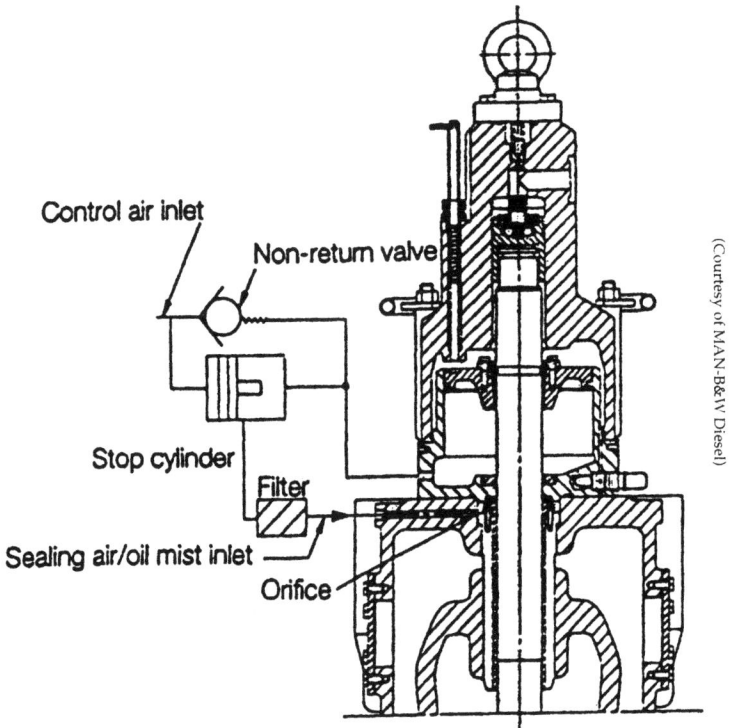

Figure 26. MAN-B&W MC Engine Exhaust Valve Sealing

On smaller engines the valves can be ground in against the valve seat using carborundum grinding paste. Use coarse paste first until all blemishes have been removed and then finish with medium then fine paste, until the seal faces have a smooth finish. After grinding ensure all components are thoroughly washed in kerosene and cleaned to remove all traces of grinding paste. Lubricate the valve stem before assembly when refitting the valves.

Grinding paste should not be used on larger valves. These should be ground with a special grinding machine set at the correct seat angle. The valve seat does not have to be ground unless a new or reconditioned valve seat is being fitted or the seat has been damaged. After grinding valves or seats all parts should be thoroughly cleaned to remove any grinding dust.

To check the valve will seat properly after grinding the valve face should be 'blued' and pressed against the seat. If the angle is correct there should only be a narrow blue band on the seat, rather than being blue all over. The differential angle between seat and valve ensures full sealing when the valve is at working temperature. There is also a limit to how much a valve can be ground. Some engine manufacturers provide templates to ensure the valve face is within limits.

Valve cage cooling water spaces should be checked for any sign of deposits or corrosion which could suggest inadequate water treatment.

2.2.3   Pistons

The piston is usually of composite construction consisting of a piston crown and skirt. Modern crosshead engines have oil cooled pistons, with the oil supply and return via the piston rod.

The crown is typically forged from a heat resistant steel, such as chrome, molybdenum, nickel alloy steel. This has a higher tensile strength and good crack resistance. The resistance to wear is not so good and therefore ring grooves are usually chrome plated or hardened.

Typically four to six piston rings are fitted. The top ring is usually a harder material than the rest.

On older loop scavenged crosshead engines the piston is fitted with a long skirt, which is used to seal the scavenge and exhaust ports when the piston is at top dead centre. The skirt is usually cast iron as the thermal and mechanical loads on the skirt are much lower than on the crown. Brass rubbing bands may also be fitted. Modern long stroke engine pistons are not fitted with skirts as there are only scavenge ports in the liner. For trunk piston engines the construction and materials depend on the size of engine, type of fuel and engine speed. Small engines may have a one piece casting. This is usually aluminium alloy which has good thermal conductivity. Although cast iron has poor resistance to creep, fatigue and cracking it can be used on smaller engines where the section is small and thermal stresses lower.

Larger engines may have a two piece piston, particularly those designed to burn heavy residual grade fuel. The crown would be forged heat resistant steel with a cast iron or aluminium alloy skirt.

Piston Cooling

The piston is subject to direct flame temperature, however, the material must be kept below 500°C to maintain its strength. The piston crown must be of a thick enough section to transmit the gas loads to the rod. Heat flows must be symmetrical as overheating the piston crown can cause excessive distortion which alters the ring/liner profile causing increased wear. In extreme cases this can cause piston seizure.

There must always be adequate clearance between piston and liner to allow for expansion. Some pistons are slightly tapered above the top ring to accommodate this expansion. The amount of clearance varies depending upon the engine rating and piston material. It may be as low as 0.01 per cent bore diameter for a large diameter forged crown, to as much as 0.1 per cent bore diameter for an aluminium alloy piston. (See Figure 27.)

Aluminium alloy has very good thermal conductivity and heat will be conducted away rapidly from the crown, down the skirt, through to the liner.

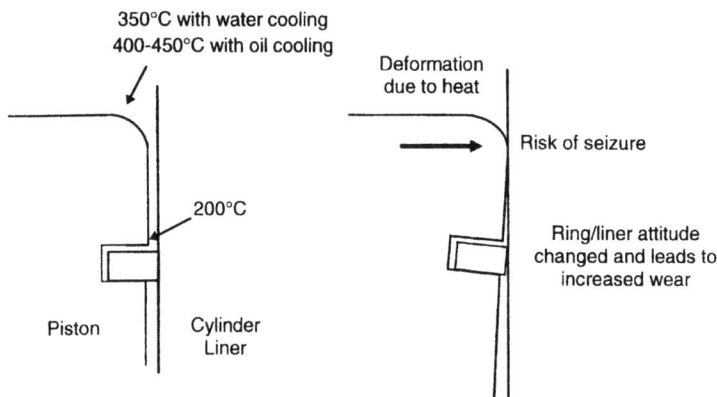

Figure 27. Distortion of Piston at High Temperatures

On medium speed engines piston cooling is performed by the crankcase system oil which is supplied via the connecting rod. Oil passes up the connecting rod to lubricate the gudgeon pin. From here the oil can either pass out of the top of the connecting rod, spraying onto the piston underside, or pass through galleries into a cooling space cast into the piston.

Where two piece pistons are used the oil will pass into a cooling space underneath the crown. The oil usually enters at a higher point than it leaves so the cooling space is not completely filled. This so called 'cocktail shaker' action improves cooling.

Figures 28, 29 and 30 show different medium speed engine pistons. MAN-B&W use a forged steel bore cooled crown and nodular cast iron skirt. The oil for cooling flows up the connecting rod into the piston crown, passes through the bores into the outer space and then drains to the crankcase. This ensures the piston cooling space is not completely full of oil and also generates a 'cocktail shaker' action.

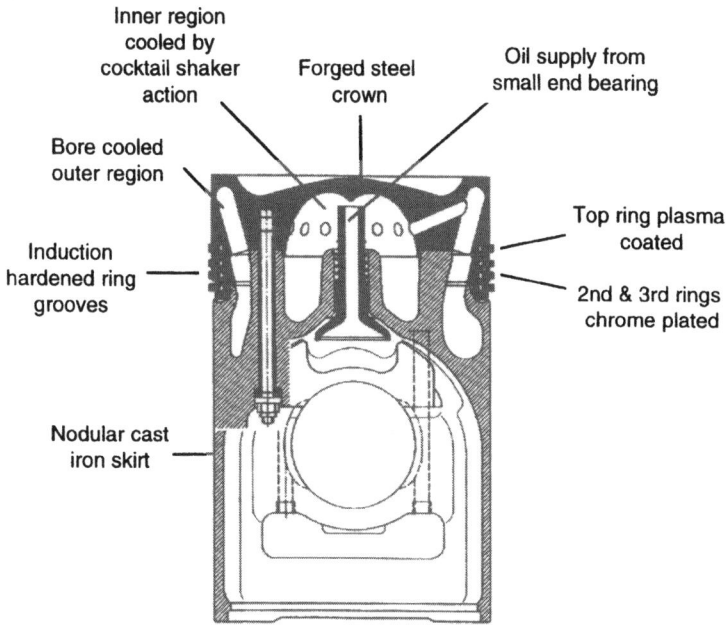

Figure 28. MAN-B&W V48/60 Piston

A unique feature of the Sulzer ZAS piston, as shown in Figure 29, is the rotating piston common to all their 'Z' engines. The spherical connecting rod is provided with a set of pawls which rotate the piston a small amount with each swing of the connecting rod. Advantages claimed for this system are:

- With each stroke a new oil wetted part of the piston comes into contact with the loaded region of the liner, reducing the risk of seizure.

- As the piston rings also rotate, the ring gap is not in the same position. This reduces local overheating of the liner due to blowby through the ring gap.

- The piston is symmetrical, therefore, thermal and mechanical deformation is symmetrical.

- Lubricating oil is evenly distributed.

- Smaller piston/liner clearances are possible which reduces piston rocking.

- The oil scraper ring can be placed at the bottom of the skirt, giving better oil control.

- Reduced lubricating oil consumption.

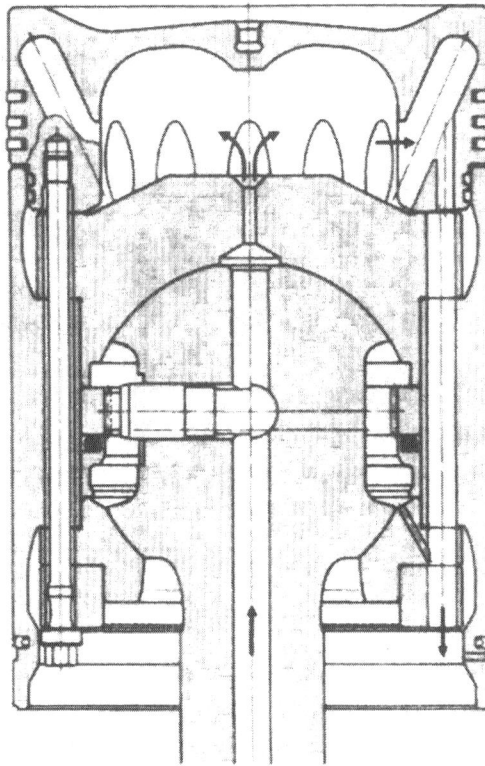

Figure 29. Wärtsilä NSD ZAS Piston

Figure 30 shows a Wärtsilä piston with forced skirt lubrication. In this case oil gets fed under pressure into a circumferential groove at the upper end of the piston skirt.

By using forced piston skirt lubrication it has been possible to reduce the number of compression rings. Other advantages claimed are:

- More even distribution of lubricating oil over the liner surface.

- Piston rocking is reduced due to the increased oil damping effect.

- Blowby is reduced.

- Corrosive wear is reduced as the alkaline additives in the oil are replenished faster.

- Piston ring fouling is reduced as the detergent additive in the oil is replenished faster.

- Abrasive wear is reduced as hard particles are transported away by the oil.
- Improved cylinder pre-lubrication prior to starting the engine.
- Oil consumption is reduced due to the highly efficient oil scraper ring.

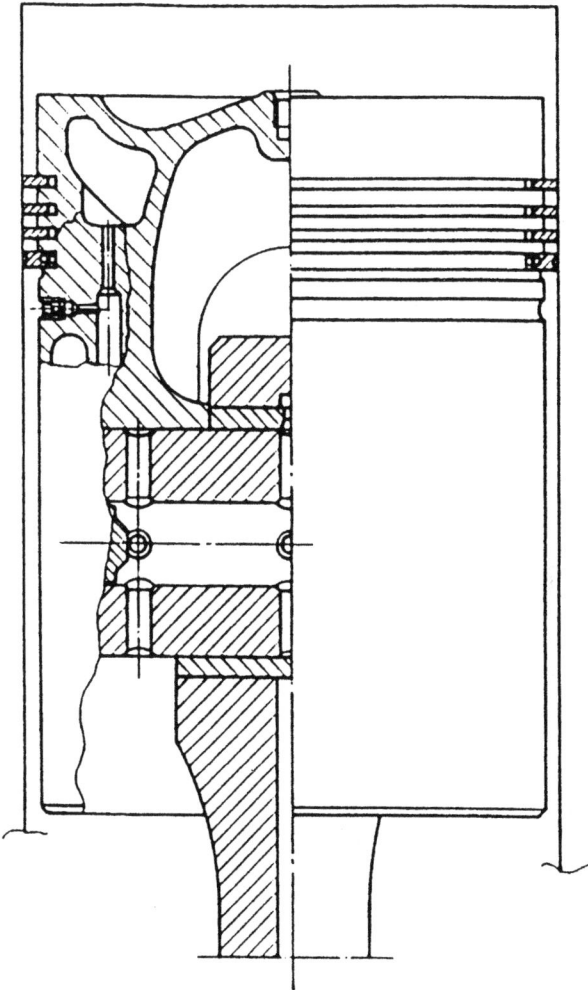

(Courtesy of Wärtsilä NSD)

Figure 30. Wärtsilä 46 Piston with Forced Skirt Lubrication

All modern crosshead engines now utilise the crankcase system oil as the cooling medium. As mentioned, oil passes up the piston rod into the cooling space. The inlet is above the outlet which gives a 'cocktail shaker' action. This can be seen in Figure 31.

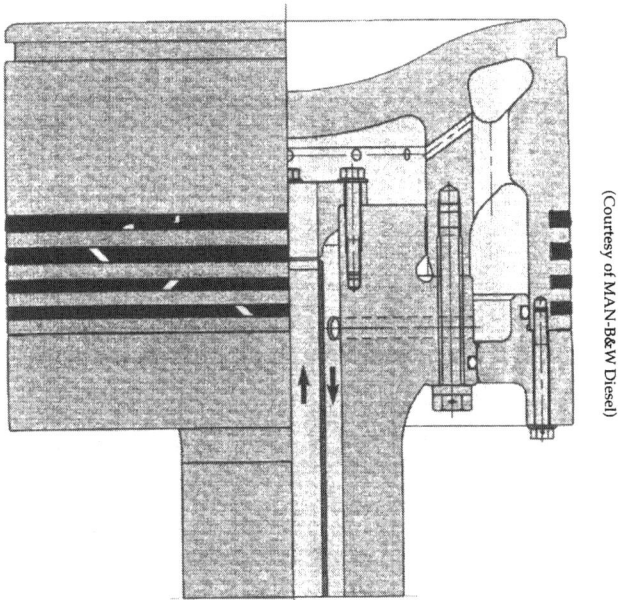

Figure 31. MAN-B&W S-MC-C Piston

Operation with defective cylinder

Operating with a defective cylinder cover where the running gear has been left in place has been described, however, problems with the running gear could mean one cylinder has to be isolated. Wherever possible the defective parts should be replaced, however, this is not always be possible and the engine would therefore have to operate with one or more cylinders out of operation. The extent of the work to be carried out depends on the nature of the problem. Table 1 shows different cases for a MAN-B&W MC engine.

For all of Table 1, except Case A, the tightness of any blanking connections or locking devices must be checked before starting the engine. The engine should be operated for a few minutes then the parts checked again for security. Oil flow to bearings should also be checked.

	Case A	Case B	Case C	Case D	Case E
Nature of emergency action	Combustion to be stopped	Compression and combustion to be stopped	Faulty exhaust valve	All reciprocating parts suspended or out	All reciprocating parts out
Possible reasons	Piston ring or exhaust valve blow-by. Reduce bearing load. Faulty injection equipment	Leaking cylinder cover or liner	Malfunction of exhaust valve or control gear	Quickest and safest measure in the event of faults in large moving parts, cylinder cover or liner	Spare parts are not available on board
Fuel pump with roller guide	Lifted	Lifted	Lifted	Lifted	Lifted
Exhaust valve	Working	Held open	Closed	Closed	Closed
Air for air spring	Open	Closed	Open	Open	Open
Exhaust valve actuator with roller guide	Working	Out or lifted	Out or lifted	Out or lifted	Out or lifted
Oil inlet for actuator	Open	Pipe dismantled and blocked	Open	Open	Open
Starting valve	Working	Blanked	Working	Blanked	Blanked
Piston with rod	Moving	Moving	Moving	Suspended	Out
Crosshead	Moving	Moving	Moving	Suspended	Out
Connecting rod	Moving	Moving	Moving	Out	Out
Crankpin bearing	Moving	Moving	Moving	Out	Out
Oil inlet to crosshead	Open	Open	Open	Blanked	Blanked
Cooling oil outlet from crosshead	Open	Open	Open		
Cylinder lubricator	Working	Working	Working	Zero delivery	Zero delivery

Table 1. Engine Operation with Defective Running Gear

The engine must be operated at reduced speed and closely monitored for excessive temperatures and vibration. The engine 'breathing' will have changed if a cylinder is not operating and exhaust temperatures on other cylinders may increase, resulting in uneven exhaust temperature.

In addition vibration may increase and any previous barred speed ranges may have changed, particularly if any of the running gear have been removed. The engine manufacturer should always be contacted if this situation arises. They can then calculate a new barred speed range.

Piston Rings

Piston rings are an important part of the engine yet they tend to be neglected because they are a consumable item. Often they can be found in the corner of a storeroom covered in dirt and rust, and in severe cases they have been painted.

Piston rings should be kept in a dry place and stored flat. A light coating of a protective lubricant will prevent them rusting. If they are covered, make sure they are checked regularly to ensure they are in satisfactory condition.

The function of a piston ring is to:

- Provide a seal to the combustion chamber to prevent gases and combustion products passing the piston.
- Control the lubricating oil.
- Conduct heat away from the piston to the liner.

Generally, there are two types of piston ring currently in use:

- *Compression ring.* Provides a gas seal.

- *Scraper or Oil Control ring.* Distributes oil on the cylinder liner preventing the oil passing upwards into the combustion chamber. These rings are normally found on trunk piston engines.

The piston ring sits in a machined groove, located such that the ring operates at an acceptable temperature. If the rings were fitted too high, the high temperatures would rapidly burn off the oil and the rings would seize in their grooves. The piston ring must be free to move in its groove, therefore, a clearance is required. Ring clearances can be summarised as follows:

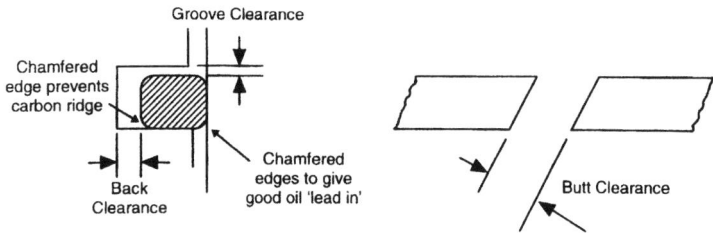

Figure 32. Piston Ring Clearances

Groove clearance	Allows pressure to build up behind the ring. Allows oil to flow into the groove. *Too small* - Ring will stick in groove. This will result in poor sealing and possible blowby which will burn away the oil and cause scuffing. Insufficient gas pressure behind the ring will affect sealing. *Too large* - Ring flutter and possible breakage.
Back clearance	Allows pressure to build up behind ring.
Butt clearance	This may also be termed gap clearance and is required to accommodate the ring expansion as it heats up. *Too small* - As the ring expands the butts will come together. This will exert a large radial pressure on the liner, breaking down the oil film and increasing scuffing wear. Ring seizure may occur. *Too large* - Excessive gas leakage.

The piston rings operate in a hostile environment. The load is fluctuating. and at top dead centre the rings are at their slowest speed and highest temperature. The rings must withstand corrosive combustion products. Piston rings must therefore have high tensile strength to resist breakage, combined with good anti-corrosive properties. Rings must also maintain tension at lower combustion pressures and be compatible with the liner material.

Achieving high strength and good resistance to wear is not easy, therefore, a high strength material is generally used. A surface treatment can be used if anti-wear properties are required.

Piston rings are usually manufactured from flake graphite alloy cast iron. A typical composition for a modern two stroke piston ring is as follows:

Element	Symbol	%
Carbon	C	3.15
Silicon	Si	1.55
Phosphorus	P	0.20
Manganese	Mn	0.90
Chromium	Cr	0.15
Molybdenum	Mo	0.60
Copper	Cu	0.75

Surface treatments

Surface treatments fall into two categories:

1) Those which are expendable and used to resist corrosion while in storage and to assist the running in process. This surface treatment is usually achieved by chemical processes which produce oxides or phosphates of various metals such as iron, zinc and manganese. Copper and chrome plating may also be used.

2) Those which improve the wear resistance of the ring for its operating life. This is usually achieved with chrome plating, flame hardening or plasma spraying. Chrome plating is perhaps the most common and is often seen on four stroke engine piston rings. Ring grooves can also be chrome plated. Chrome plating provides a very smooth surface which can help prevent the build up of deposits.

Some piston rings are nitrided to give increased hardness as well as improved wear and fatigue resistance. As the surface is porous it tends to retain oil and sulphur compounds which can act as a solid lubricant.

Piston rings are either cast as single pieces or machined from a cast cylinder or 'pot'. Traditionally, rings are pretensioned to achieve the required wall pressure. This is done by hammering the inner surface, starting opposite the gap then decreasing the hammer force as the hammer moves round to the gap. The ring tension holds the ring against the liner wall at low engine speeds when the combustion pressure is lower.

A modern method of manufacture is cam turning. This is where the ring is cast in the unfitted shape and then machined to give the required ovality and ring pressure. Figure 33 shows different ring profiles and distributions of wall pressure.

(Courtesy of Standard Piston Ring Co Ltd)

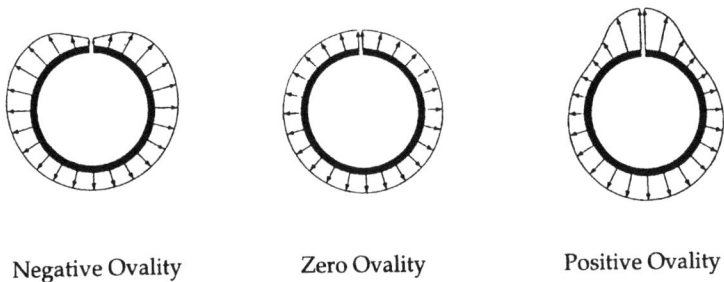

| Negative Ovality | Zero Ovality | Positive Ovality |

Figure 33. Piston Ring Wall Pressure Distributions

Negative ovality rings are typically used in two stroke engines and have reduced wall pressure at the ring ends. In contrast a high performance four stroke engine would have positive ovality rings, which dampen the ring ends to prevent them fluttering at high speeds. Zero ovality has constant wall pressure around its circumference and is normal for most types of piston ring.

Piston ring maintenance and problems

Piston ring sticking may be a problem, however, it is possible to check if the rings are free during an inspection through the scavenge ports. A piece of flat wood or doweling can be used to press the rings and see if there is tension in the ring. Lack of tension indicates a possible broken ring and a dark or black appearance may indicate blowby. It may be possible to see breaks in the ring or there may be large gaps where sections of ring have broken. Broken pieces of ring usually end up in the exhaust manifold, which should also be inspected.

The causes of broken rings may be due to striking hard edges, such as ports, excessive wear of rings and/or grooves, allowing the rings to tilt slightly, poor quality fuel leading to excessive peak pressures (high CCAI - see section 2.4) or collapse of the piston ring. Heavy starting, which results in excessive cylinder pressures which often lift the cylinder relief valve, can also lead to ring breakage.

Ring collapse is probably the main cause of ring failure. This is due to insufficient gas pressure behind the ring. If the ring is not exerting sufficient pressure on the liner wall gas pressure will penetrate between the ring and liner causing the ring to collapse into the groove. This will eventually result in breakage. Causes of ring collapse are:

- Deposits in the ring groove.
- Insufficient ring clearance.
- Rings sticking in grooves.
- Poor sealing between the ring and ring groove lower face.
- Excessive chamfering on ring butts or ring edges.
- Cloverleafing of liner.

Besides a risk of ring breakage, worn rings can lead to blowby and a risk of scavenge fires. Blowby will destroy the oil film and lead to an increased wear rate. The localised overheating may also cause a cracked liner. Blowby is usually indicated by darkened areas on the ring.

During piston overhaul all the ring grooves should be thoroughly cleaned and the groove and gap clearances checked. The condition of the rings and grooves should be noted. If any rings are broken the location of the break, i.e. 'opposite ring gap' or 'at ring gap', should also be noted. Ring surfaces should normally be smooth but may also show signs of scuffing or abrasive wear.

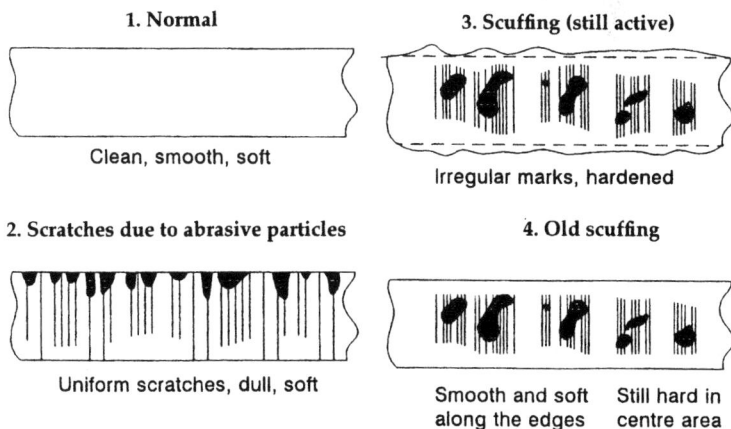

1. Normal

Clean, smooth, soft

2. Scratches due to abrasive particles

Uniform scratches, dull, soft

3. Scuffing (still active)

Irregular marks, hardened

4. Old scuffing

Smooth and soft   Still hard in
along the edges   centre area

(Courtesy of MAN-B&W Diesel)

Figure 34. Piston Ring Surface Condition

Ring gaps should be checked by fitting them in the liner at their normal top dead centre position. Rings should be fitted alternately between left and right hand cuts, ensuring the gaps are spaced 180° apart (as shown in Figure 35).

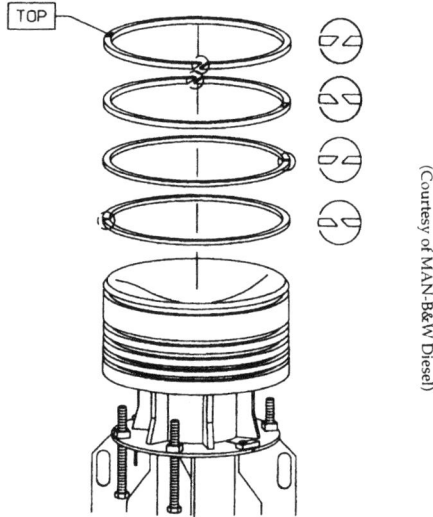

(Courtesy of MAN-B&W Diesel)

Figure 35. Fitting Piston Rings

Some rings, particularly those that are chrome plated, may have 'Top' marked on the upper surface. Rings should be checked for any such markings and then fitted correctly. When fitting larger rings a special ring tensioner is usually provided which applies uniform tension to the ring and prevents deformation. It also does not unduly stress the ring. The ring tensioner should also be used for ring removal.

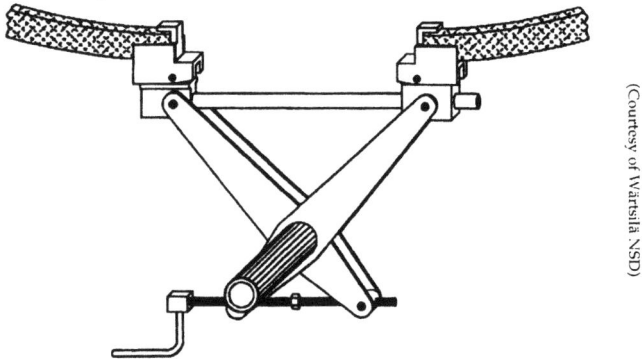

(Courtesy of Wärtsilä NSD)

Figure 36. Tool for Removal and Fitting Of Rings

Prior to refitting a piston a special guide ring - usually supplied by the engine manufacturer - should be placed on top of the liner so the piston rings are compressed as the piston is lowered into the liner. This operation should be carried out slowly and the rings should be visually checked to ensure they compress as the liner lowers. If a ring jams then the whole piston assembly will be supported by the jammed ring. The ring pack should be liberally lubricated before lowering the piston into the liner.

If replacing a few rings bear in mind that, although the rings may be within limits, they may have to operate for another 10 000 hours or more. By this stage the rings may be worn to the point where they begin to break. The loading on the engine and type of fuel being used should also be considered. Rings are a relatively cheap item and it is much easier to renew the rings during piston overhaul than in an emergency.

2.2.4 Cylinder Liners

The cylinder liner needs to be able to withstand two main stresses:

1. Hoop stress due to gas pressure.

2. Thermal stress due to the temperature difference between the combustion chamber and cooling space.

The stresses set up in a liner are complex. On the cooling water side the surface will be in tension so this is where any cracks will initiate. The crack will then propagate through the liner material until it reaches a region of compressive stress. The crack will then move circumferentially.

Vertical cracks in the liner are usually due to creep because the material has been overheated. Overheating can be due to several causes.

- Lack of cooling caused by events such as a loss of coolant flow, heavy scaling of cooling surfaces or a fouled cooler.

- Combustion problems such as flame impingement. This leads to localised overheating due to faulty fuel injectors or poor quality fuel.

- Afterburning which is caused by poor quality fuel.

- Broken or worn piston rings which lead to blowby.

- Water entrainment in scavenge air which washes off the liner lubricating oil film.

In order to withstand the high gas pressures the liner needs to be of sufficient thickness. If it is too thick, however, it will lead to a higher temperature gradient which will increase the thermal stress.

Large bore liners fitted to slow speed engines, and some of the larger medium speed engines, have a thicker top section to give adequate strength. This region is bore cooled to reduce thermal stresses. The bores are usually drilled, with plugs where appropriate.

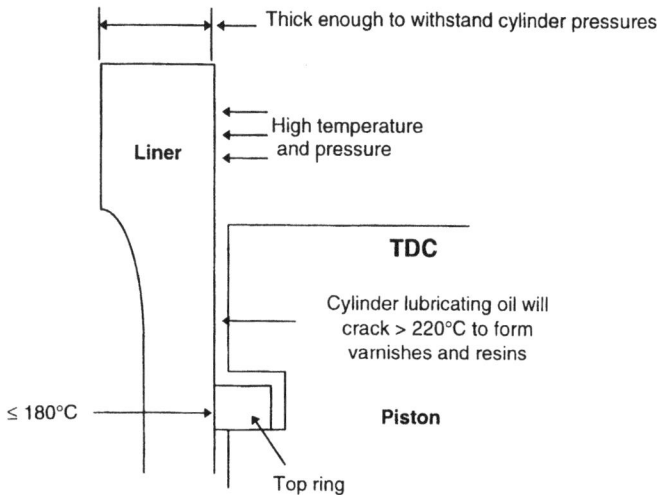

Figure 37. Liner Requirements

Figure 38 shows the application of bore cooling on Wärtsilä NSD two stroke engines. This was applied to keep stresses within acceptable limits as combustion pressures and temperatures increased.

Liner Material

As well as designing the liner section to give good strength, the choice of material is a most important factor. Liners are usually grey cast iron with small additions of nickel, chromium and molybdenum. The exact microstructure is controlled by the rate of cooling and the alloying elements, however, it is usually a flake graphite cast iron with pearlitic matrix.

Engine manufacturers usually have their own liner material specifications and there have been many changes to liner materials due to the increased demand for higher outputs.

Wärtsilä NSD RTA: Lamellar or vemicullar graphite cast iron.

Wärtsilä NSD 32/46:   Grey cast iron alloy.

MAN-B&W MC: Tarkalloy (a lamellar graphite cast iron with boron and phosphorus. Tarkalloy is a brand name).

Dual layer liners may be used where a cast steel liner has Tarkalloy centrifugally cast onto it. The cast steel gives high strength and Tarkalloy provides good wear resistance. Due to the high cost of these liners they are usually in two parts. Many engines now incorporate anti-polishing rings, fitted at the top of the liner. These are slightly smaller than the liner bore and help to prevent carbon deposits on the top land of the piston. These deposits can damage the liner and affect the lubricant film.

Cylinder Wear

During its working life the cylinder liner will gradually wear. On a crosshead engine this is usually of the order 0.10mm per 1000 hours. On medium speed engines it is much lower, usually 0.015mm per 1000 hours. Liner wear is either corrosive, abrasive or scuffing. All three may be occurring at the same time, although at very low levels.

Corrosive wear

This is caused by the acids which are formed during combustion. In modern engines this is very low due to a combination of wear resistant materials and, principally, highly alkaline crankcase and cylinder oils with the ability to neutralise combustion acids. If water is present in the fuel or charge air then excessive amounts of acid can be formed which may not be neutralised effectively. It is therefore important to ensure the charge air is well drained.

The liner must also be kept at the correct temperature. A 'cold' liner wall will cause acids to condense once the temperature falls below the acid dew point. If this occurs, maximum corrosion will be at the top of the liner where the acid is hottest.

RN
1968

RN-M
1975

RL
1977

RTA
1982

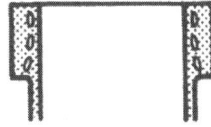
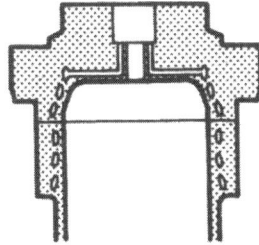
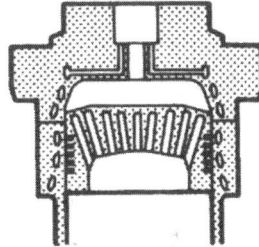
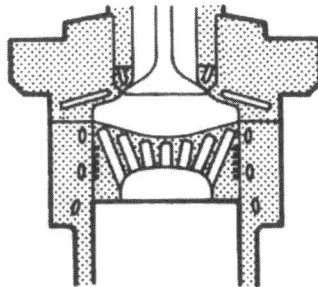

(Courtesy of Wärtsilä NSD)

Figure 38. Application of Bore Cooling in Wärtsilä NSD Two Stroke Engines

'Cloverleafing' is a term used to describe longitudinal corrosive wear between the lubricator ports due to the breakdown of the oil film, caused either by water in the charge air, or flame impingement from a faulty fuel injector, which will burn off the oil.

Corrosive wear can be identified by a surface that is pock marked with tiny craters. This is where acid has eaten into the metal.

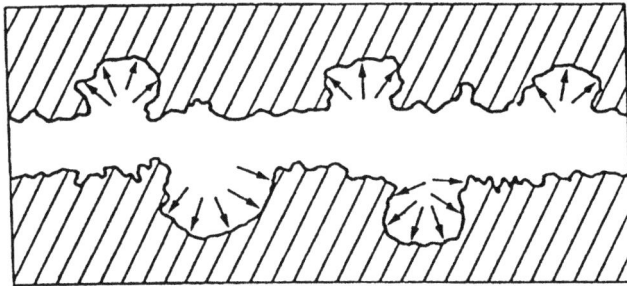

(Courtesy of Standard Piston Ring Co Ltd)

Figure 39. Schematic Diagram of Corrosive Wear

Abrasive wear

Abrasive wear is caused by hard carbonaceous combustion products, dirt in the scavenge air or abrasive particles in the fuel. The hard particles score the liner and rings. These particles can be reduced by ensuring the fuel is efficiently centrifuged and filtered, and that turbocharger filters are in good condition. The abrasive particles cause the liner and ring faces to have fine vertical score marks. Abrasive wear can also occur as a side effect of other wear processes.

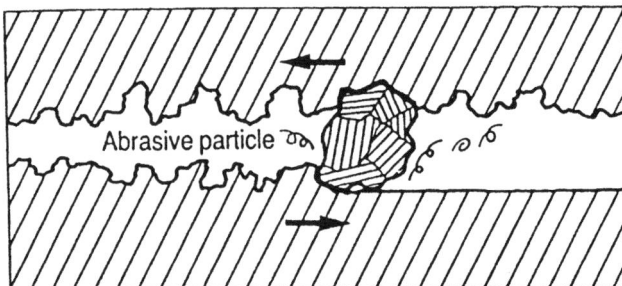

Abrasive particle

(Courtesy of Standard Piston Ring Co Ltd)

Figure 40. Schematic Diagram of Abrasive Wear

Scuffing wear

Scuffing occurs when metal to metal contact takes place. In cases of severe wear the rate may be as high as 2-3mm per 1000 hours causing permanent damage to the liner. The maximum oil film thickness between the liner and rings is usually about 5 microns. Fluid film lubrication will occur at mid stroke, when the piston speed is at a maximum. At top and bottom dead centre, when the piston speed is zero, boundary lubrication prevails. Under certain circumstances it may not be possible to maintain an adequate film thickness. Metal to metal contact will then occur.

Wear has a transient nature with two well defined regions, mild and severe wear. The change from one to the other is very rapid.

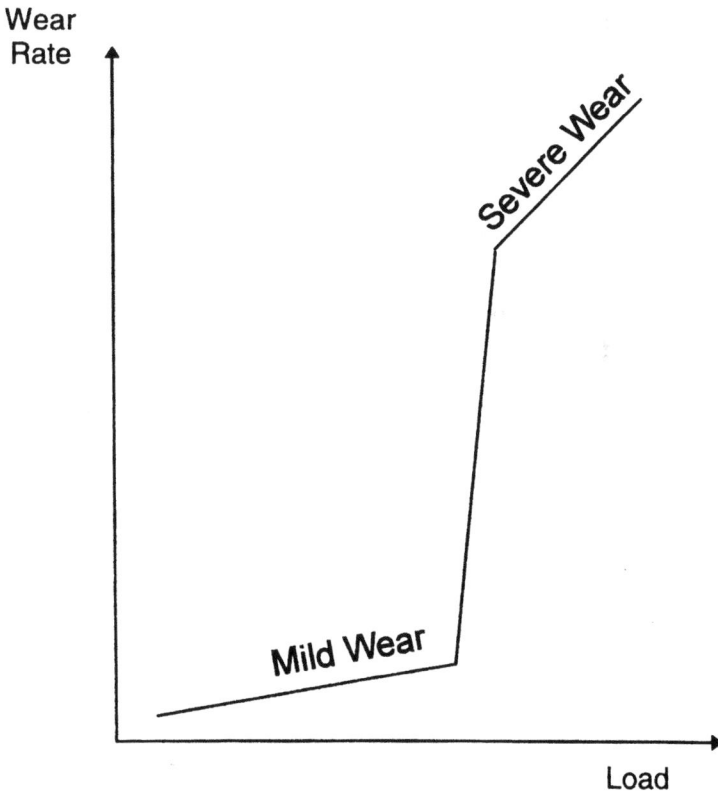

Figure 41. Mild and Severe Wear

During the mild stages of wear the debris consists mainly of iron oxides. If metal to metal contact occurs the material will quickly overheat. Surface asperities will weld together (micro seizure) and are then torn out of the parent metal. The high temperature generated reduces the fatigue resistance of the cast iron and the fluctuating pressure causes cracks to form below the liner surface. Small particles of metal, which have been hardened by the high temperature, will flake off causing a severe wear situation to develop. Scuffing can occur due to the following:

- Lack of, or insufficient, lubricant.
- Incorrect cylinder oil timing.
- Blowby, which destroys the oil film.
- Water entrainment in scavenge air, washing off liner lubricating oil film.
- High ring/liner pressure due to increased combustion pressure. This may be due to poor quality fuel, incorrect injection timing or overload.
- Worn liner or rings. Worn ring grooves allow the ring to tilt which will break down the oil film.

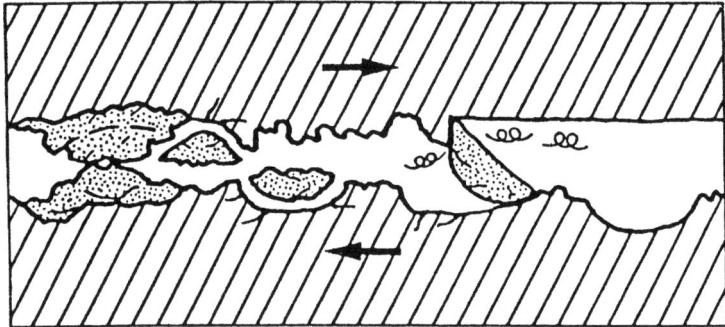

(Courtesy of Standard Piston Ring Co Ltd)

Figure 42. Schematic Diagram of Scuffing Wear

Liner lubrication

Crosshead engine cylinder oils need to perform a different duty to the crankcase oil and are formulated with a different additive package. The main functions of a cylinder lubricant are to:

- Maintain a gas seal between ring and liner.

- Provide a transport medium for the alkaline additive.
- Lubricate the ring pack.
- Carry combustion products and wear debris from the liner surface.
- Reduce sliding friction.
- Prevent the build up of deposits.

In modern crosshead engines the oil has to work in a hostile environment. The trend for longer piston strokes leads to the oil being exposed to high temperatures for longer periods.

Sufficient oil must be injected to ensure adequate oil film thickness. This relatively small quantity of oil must quickly cover the liner surface. Good spreadabilty is therefore desirable. The position of the lubricator ports is important. If positioned too high the oil will be exposed to excessive temperatures and will quickly burn away. There must also be a sufficient number of ports to ensure adequate oil film around the liner.

Generally there are upper and lower ports. The upper ports are set to inject oil between the top two or three rings.

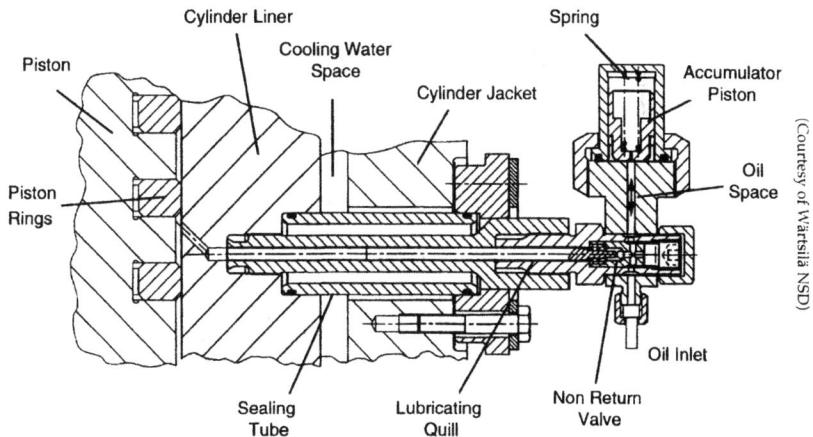

Figure 43. Cylinder Liner Lubricator

(Courtesy of Wärtsilä NSD)

Liners in trunk piston engines have traditionally been lubricated by oil splashing up from the crankcase. This is still the case in many engines, however, some of the larger engines now incorporate direct lubrication systems, such as bores in the liner to supply clean lubricating oil, or forced skirt lubrication via the piston cooling system (see Figure 30).

If an engine has been stopped for any period of time the liner may become dry. Starting the engine will lead to increased wear until the oil film has been re-established. Prior to starting the engine, it should be turned and the lubricators operated. On some engines, where the load may be rapidly applied, the pre-lubrication may be automatic. The engine will be slowly turned while a metered amount of oil is supplied to keep the liner wet.

Liner Inspection and Measurement

Before the piston can be removed the wear ridge that forms on the liner must be ground smooth. This ridge forms at the top dead centre position of the top piston ring. Once the piston has been removed the liner should be cleaned and visually inspected for any signs of scuffing, scoring or corrosive wear. Dark areas on the liner usually indicate blowby caused by poor ring sealing or broken rings. As the liner wears the lubricator grooves and scavenge ports may form sharp edges. These should be removed with an oil stone. The depth of the lubricator grooves should also be checked against the manufacturers' limits.

The liner should be calibrated to check for wear and ovality. If the liner is still warm the calibration gauge should be allowed to warm up to a similar temperature by placing it on an adjacent cylinder head. If a guide is not available for taking measurements in the correct position, one can easily be made from a thin strip of steel drilled at the appropriate positions for the gauge.

Gauge readings should be taken, fore and aft and port and starboard to check ovality. Any undue uneven wear of the liner can be tested by fitting a new piston ring in the liner. Gauge readings should be taken at points identified by the engine manufacturer, usually three or four at the top of the running region then spaced out at greater intervals further down the liner. All measurements should be recorded for future reference.

When removing a piston for overhaul the following information should be recorded: ring clearances; number of rings replaced; condition of piston crown and grooves; condition of liner and ports; all liner calibrations; all relevant information such as engine hours, hours since last overhaul, etc.

On a medium speed engine the cylinder liner can become highly polished which leads to an increase in oil consumption. When this happens the liner must be honed with a special honing tool. This tool consists of a set of oil stones that rotate at a fixed speed while moving up and down the liner.

Running in Cylinder Liners

The 'running in' process needs to be gradual to ensure the rings are well bedded in. A certain amount of wear is desirable so that sealing between the ring and liner is achieved quickly. In a two stroke engine a low alkaline cylinder oil should be used to increase corrosive wear, however, this is not always available. Crankcase oil should not be used as it has a much lower viscosity than cylinder oil. The oil film thickness may be inadequate which could lead to scuffing and permanent damage to the liner. Ideally a couple of drums of low alkaline SAE 50 oil should be kept on board for running in liners. If this is not available then normal cylinder oil should be used.

Whilst running in the engine the load must be gradually applied as per the manufacturer's guidelines and the cylinder oil feed rate may have to be adjusted.

Figure 44 shows a typical running in programme for a MAN-B&W two stroke engine.

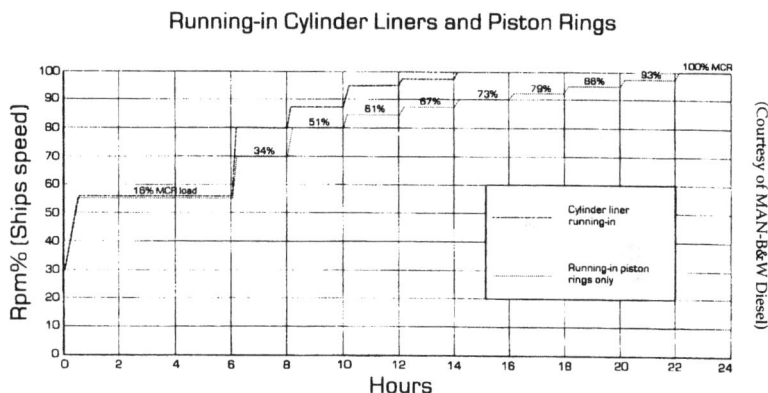

Figure 44. Running-in Liners and Piston Rings

2.2.5   Fuel injectors

The fuel injector essentially consists of a nozzle body with a nozzle and needle, which are generally cooled in larger engines. A compression spring together with a spring adjuster controls the opening pressure. The fuel injector is set to open at a predetermined pressure so that the velocity of the fuel is high enough to atomise as it is injected into the cylinder. The components of a fuel injector are shown in Figure 45.

Spring tensioner

hp fuel pipe connection
piece with circulation
valve

Compression spring

Nozzle holder or
injector body

Spring carrier

Cap nut

Push rod

Nozzle with needle

Atomiser

(Courtesy of Wärtsilä NSD)

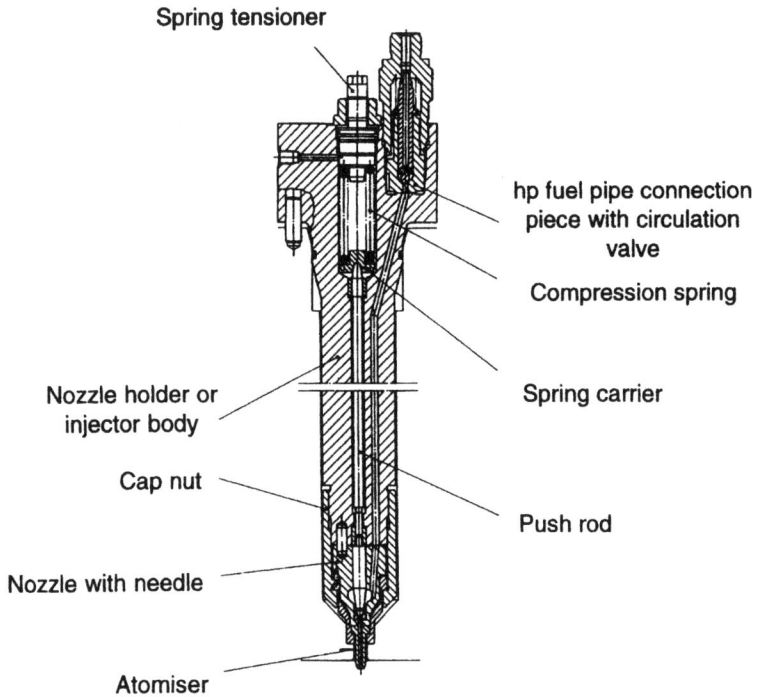

Figure 45. Wärtsilä NSD RTA Fuel Injector

Fuel flows through the injector body to the annular space around the needle seat, the pressure spring and push rod holds the needle onto its seat. During the injection stroke the pressure of the fuel rises rapidly. Once it exceeds the spring pressure the needle is lifted off its seat by the pressure of the fuel acting on the shoulder of the needle. Fuel is then injected into the cylinder.

The fuel must be well atomised and mixed with the charge air to give good combustion. Factors that play a part in this are:

- Fuel penetration.
- Air swirl.
- Number of injector holes.
- Diameter of injector holes.
- Length and angle of injector holes.

Penetration is the distance the fuel spray travels into the combustion chamber and depends on the viscosity of the fuel and its pressure. For a jerk type pump the injection pressure is a function of the number and size of nozzle holes and also the velocity of the pump plunger. The latter is affected by the engine speed.

Traditional loop scavenged crosshead engines have a central vertical fuel injector, as do trunk piston engines. Modern crosshead engines utilise uniflow scavenging with a central exhaust valve, with the charge air rotating around the cylinder vertical axis. This is known as swirl and is achieved by making the scavenge ports tangential to the cylinder bore. Swirl promotes better mixing of the fuel and air.

As the exhaust valve is located centrally in the cylinder head, one or more injectors - usually two or three - are placed to one side of the exhaust valve. Figure 46 shows the temperature distribution for a Wärtsilä NSD RTA84T engine with three injectors per cylinder.

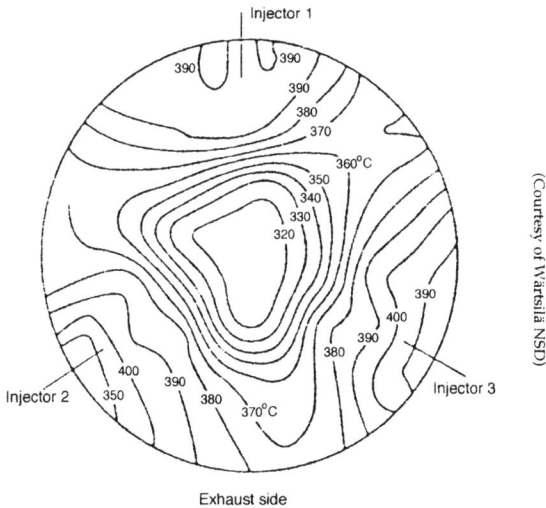

Figure 46. Piston Crown Temperature Distribution with Three Injectors

The advantages claimed for this are:

- More uniform temperature distribution around the cylinder head, piston and liner.
- More efficient combustion, therefore, better fuel economy.
- Lower exhaust valve seat temperature.

Overhaul and Testing Fuel Injectors

When overhauling injectors it is important that the work area is very clean. Lay paper out on the work bench - disused charts are ideal. Do not use rags or cloths for cleaning components. Wash all items thoroughly in kerosene and blow dry with compressed air.

Inspect all sealing faces for any damage. Small scratches can be removed by lapping in against a special surface plate, using a fine lapping paste such as jewellers rouge. Any pitting or indentations of the sealing face mean the nozzle should be replaced or sent to specialists for reconditioning.

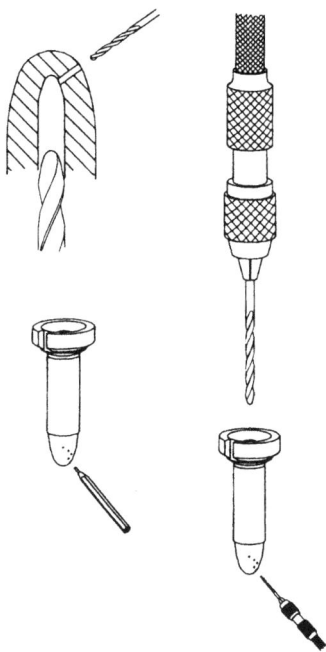

(Courtesy of MAN-B&W Diesel)

Figure 47. Cleaning Injector Nozzle

Clean the nozzle bore with a special drill and use specially supplied needles for cleaning the nozzle holes. If the holes are too large, or appear oval when inspected with a magnifying glass, the nozzle should be discarded.

The nozzle needle and guide are a matched pair and should not be interchanged. The needle should be free to move in the guide and when lifted it should drop down into the guide under its own weight.

Reassemble the injector and tighten the cap nut down to the correct torque. Note that some engine manufacturers supply different sets of nozzles for the same engine. An example of this would be slow steaming nozzles, which may have different size holes. Always make sure the correct nozzle has been fitted.

Testing Injectors

Mount the injector in its test rig and connect up the oil supply.

Note: Under no circumstances should hands be placed under the injector spray. The high velocity oil jet can penetrate the skin and cause blood poisoning.

Key

1. Oil container

2. Pressure gauge

3. Shut off valve

4. Pump lever

5. Test pump

6. Injector

7. HP fuel pump

(Courtesy of MTU)

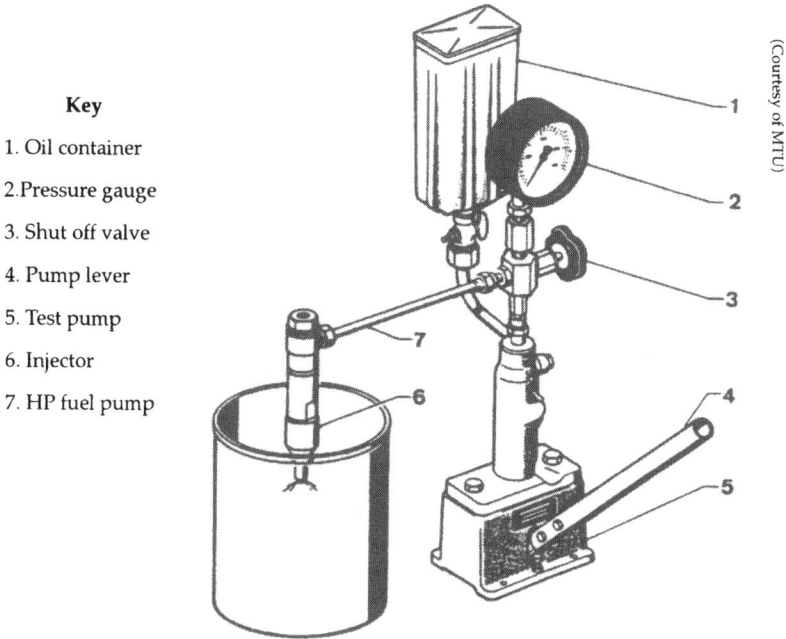

Figure 48. Fuel Injector Test Rig

With the injector priming valve open, operate the hand pump to prime the injector. Once fuel flows from the priming valve it can be closed.

Operate the pump rapidly for several strokes. The injector should open with a high pitched chatter and fuel should be emitted in a fine cloud. After the injector opens, check to make sure the pressure does not fall off too quickly.

To test for tightness between the nozzle needle and seat, operate the hand pump slowly to gradually increase the pressure until it is just below opening pressure. Maintain the pressure for a few seconds and ensure the injector is not dripping.

To test for tightness between the needle and guide, operate the hand pump to increase pressure until it is just below opening pressure. See how long it takes for the pressure to fall off. If the pressure falls quickly the needle and guide should be replaced.

Where nozzles are cooled internally these spaces should be pressure tested to check for tightness. Blank off one of the fuel valve cooling connections and fill the injector cooling space with water or fuel, depending upon the cooling medium. Then connect a low pressure air supply to the other connection. Leave the air on for a short period of time and test for internal or external leakage.

Injector Problems

Weak spring

Injection will commence before full injection pressure is reached. The fuel will be injected early and, if the pressure is too low, the fuel may not be properly atomised. This will result in an increased droplet size and slower burning of the fuel. Flame impingement could also occur.

Worn needle and guide

Leakage between the needle and guide will result in a slower build up of pressure causing the fuel to be injected later which may lead to afterburning.

Nozzle leakage

This may cause pre-ignition or afterburning if the leakage is severe.

Insufficient cooling

This can lead to carbon 'trumpet' formations on the nozzle tip. Overcooling can have the same effect.

Injector action will deteriorate over a period of time due to needle and guide wear, spring weakening and nozzle hole fouling. The injectors should be removed and the pressure tested periodically.

2.2.6   Fuel Injection Pumps

There are two basic types:

- *Helix control*: Simple and less expensive. Cavitation erosion can alter control characteristics.
- *Valve control*: More expensive. Plunger less prone to wear. Valves can be overhauled or exchanged.

A fuel pump with valve control is shown in Figure 49.

The suction valve is lifted off its seat when the plunger is at bottom dead centre. As the plunger moves upwards the suction valve will be lowered onto its seat by the lever. The spill valve remains closed. This is the point of fuel injection.

The plunger continues to rise until it reaches a point where the spill valve opens. Fuel spills back to the suction side of the pump signalling the end of injection. As the plunger moves down its stroke the pressure difference across the suction valve causes it to lift off its seat which admits fuel to the pump barrel. Further down the stroke the suction valve lifts mechanically via its pushrod and lever.

The volume of fuel delivered is controlled by the position of the spill valve eccentric shaft. This is connected to the fuel governor linkage.

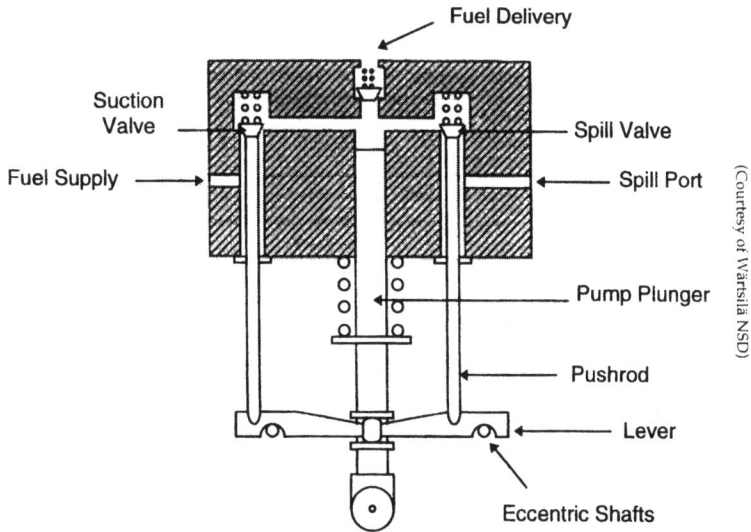

Figure 49. Schematic Diagram of Wartsila NSD Fuel Pump with Valve Control

A fuel pump with helix control is shown in Figure 50. The pump consists of a liner or barrel in which the plunger is moved up and down by the fuel cam and return spring. The suction port is in the barrel. The plunger is free to rotate in the barrel and rotation is achieved with a rack and pinion assembly.

The quantity of the fuel injected is controlled by two milled channels at the top of the plunger. Each channel has a helical control edge. Figure 51 shows how the amount of fuel injected is regulated.

Consider full fuel delivery first. In Figure 51, the plunger (1), is at the bottom of its stroke. The ports in the barrel are uncovered allowing fuel to enter the space above the plunger. As the plunger rises on the fuel cam (2), the upper edge of the machined helix covers the ports sealing off the space above the plunger. The pressure rapidly rises to injection pressure. (See Figure 52.) This is the pressure at which the injector is set to open, just before fuel is injected. Fuel pressure continues to rise as the plunger rises.

Once the lower edge of the helix uncovers the ports - (3) in Figure 51 - the fuel is vented to the pump return line and injection pressure falls. The injector will close once the pressure falls below the injector opening pressure.

The upper edge of the helix controls the commencement of fuel delivery. The lower edge controls the end of delivery.

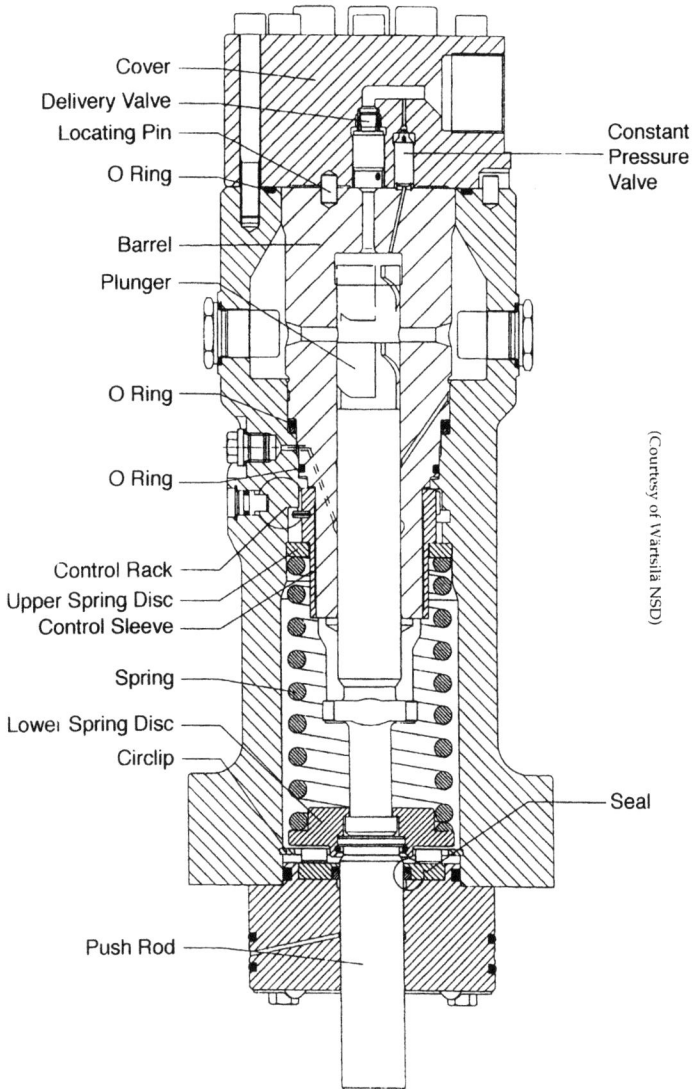

Figure 50. Fuel Pump with Helix Control

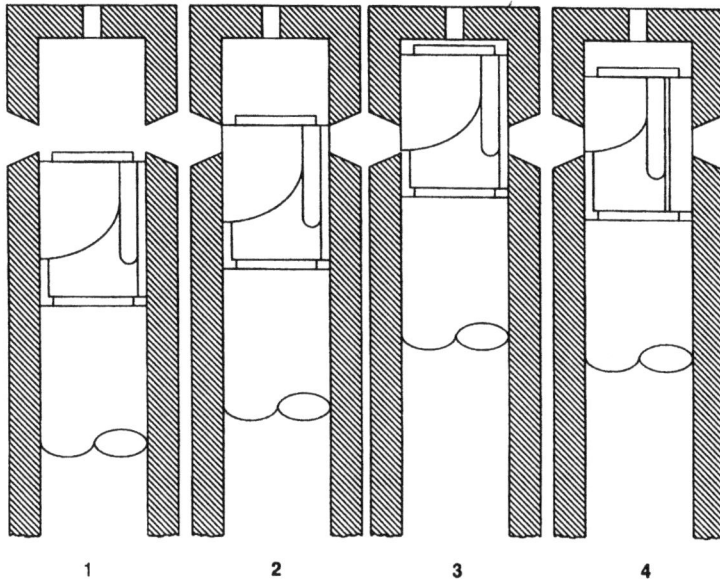

Figure 51. Regulation of Fuel Injection

Less fuel needs to be injected for part load operation. The plunger is free to rotate in the barrel via a rack and pinion assembly connected to the plunger foot. As the plunger rotates the position of the helix relative to the port in the barrel will change, as shown in Figure 51, (4). In this case the plunger does not travel so far before the helix uncovers the port, therefore, less fuel is injected.

If the plunger rotates so that the vertical milled groove is in line with one of the ports, the plunger will be at zero delivery as the pump is vented to the suction side during its entire stroke.

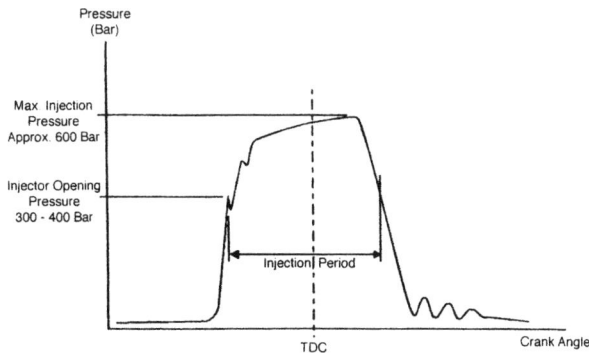

Figure 52. Fuel Pressure during Injection

Considering the plot of fuel pressure against crank angle shown in Figure 53, fuel pressure rises rapidly due to the shape of the fuel cam. When the injector opening pressure is reached, fuel injection commences. There may be a slight drop in fuel pressure at the instant the valve opens, but the pressure will continue to rise to the maximum injection pressure.

Variable Injection Timing

Variable Injection Timing (VIT) was introduced to give better fuel economy during part load operation by adjusting the beginning of injection. Timing can also be adjusted to take into account poor quality fuel. Before making any adjustments a set of draw cards should be taken to check engine compression and peak pressures (see section 2.2.7). VIT should not be used to compensate for worn cylinder components.

VIT Systems

MAN-B&W

MAN-B&W's system uses a pump barrel with a thread on the lower end. This mates with a threaded guide. Via a rack, the barrel can be raised or lowered. Raising the barrel causes the plunger to travel further before fuel is injected which retards fuel injection, and vice versa. Figure 53 shows a schematic diagram.

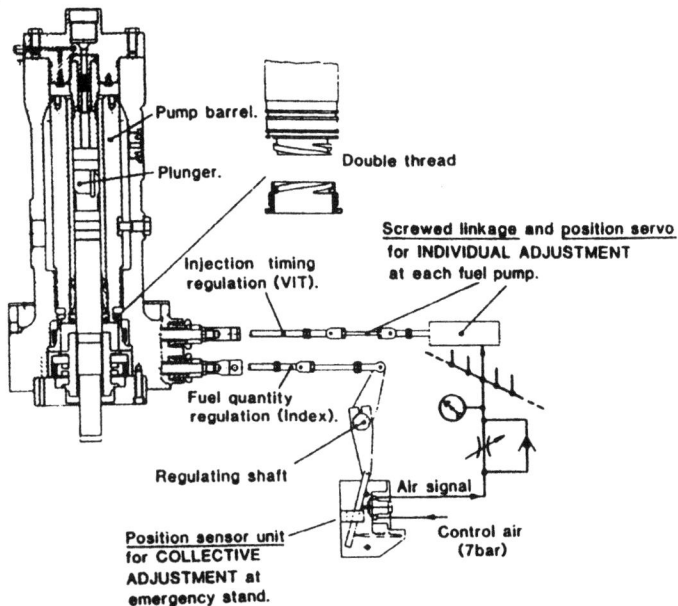

Figure 53. MAN-B&W VIT System

The VIT system gets an air signal via a position sensor unit which is actuated via the governor fuel regulating shaft. At low load the VIT is not in operation. When the power increases up to 85 per cent mcr the injection timing is gradually advanced and follows the line I - II (see Figure 54).

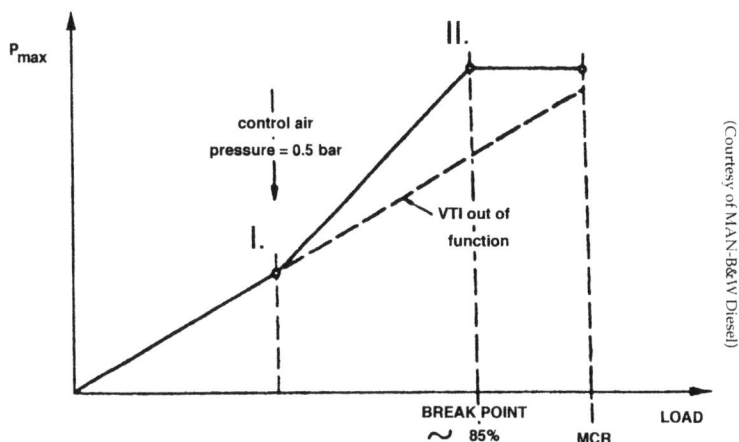

Figure 54. P_{max} Variation with Load

Above 85 per cent mcr the injection timing is retarded to maintain constant combustion pressure. The VIT linkage allows for individual adjustment of each pump.

Wärtsilä NSD RTA

The engine governor is connected via a linkage to the spill valve of each fuel pump. The spill valve determines the end of fuel injection, therefore, determining the amount of fuel delivered. The VIT system uses an electronically controlled pneumatic cylinder to control the position of the suction and spill valves simultaneously (independent of the governor). This means the beginning of injection (to control P_{max}) and duration of injection can be controlled. The control unit uses signals from the charge air pressure and engine speed as its inputs.

Increasing the VIT angle results in earlier injection which generates a higher peak pressure. Fuels of poor quality may cause reduced peak pressures. In this case there is a manually adjusted 'fuel quality setting' (FQS) which can be adjusted to increase the VIT angle. This would increase the peak pressure. FQS can be adjusted within ±2° of the VIT angle with 1° giving approximately 5 bar change in P_{max}.

Governor / Actuator

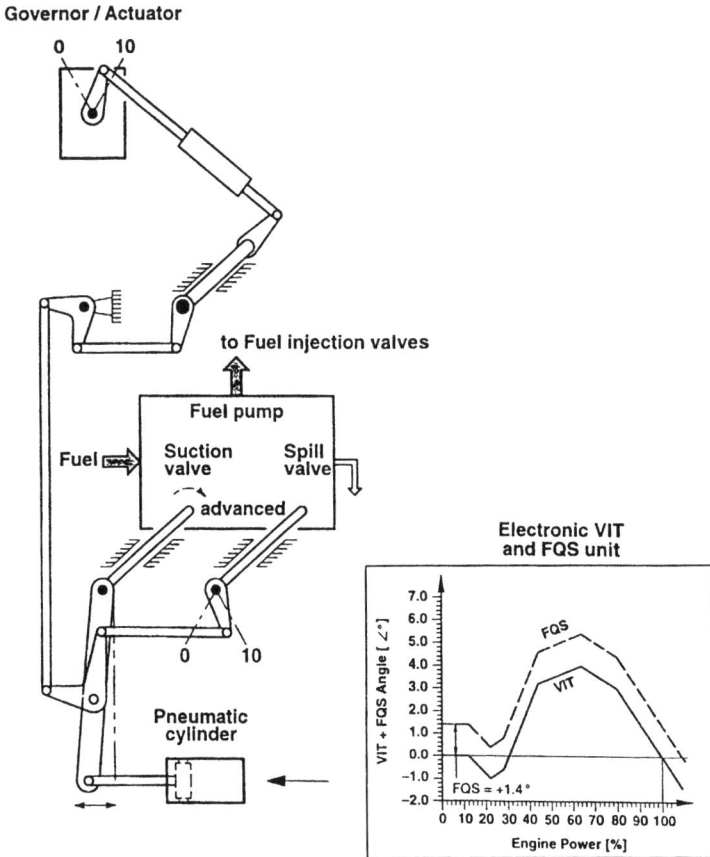

(Courtesy of Wärtsilä NSD)

Figure 55. Wärtsilä NSD VIT and FQS System

Fuel Pump Problems

The pump plunger and guide are a machined pair and the clearance between the two is very small. Over a period of time the components will wear, but the wear rate should be very low. Wear will increase if water or abrasive particles, such as catalyst fines, are present in the fuel (see Section 2.4). Wear can cause internal leakage and reduce the amount of fuel injected. If severe wear occurs there will be a loss of power on the affected cylinder.

A similar problem occurs if the fuel pump spill valve is leaking. This would mean the fuel in the pipe between the pump and injector drains out. If there is a problem that results in reduced exhaust temperature on one cylinder, the priming valve should be opened first to check whether fuel is reaching the injector. The problem may just be that the injector has become air locked. A good 'pulse' should be felt when putting a hand on the cooler sheathed part of the fuel pipe.

Engines which operate using heavy fuel oil can suffer from sticky residues gumming up fuel racks and suction and spill valves, particularly after prolonged operation at constant load. After all residues have been removed these should be freed up by working and applying a lubricant.

Fuel pumps may also leak fuel into the camshaft space. This can usually be identified with lubricating oil analysis as there will be a reduction in lubricant flashpoint together with an increase in lubricant viscosity. All fuel pump drains should be checked regularly to ensure they are clear. Figure 56 shows a fuel pump with an 'umbrella' seal to prevent contamination of the lubricating oil.

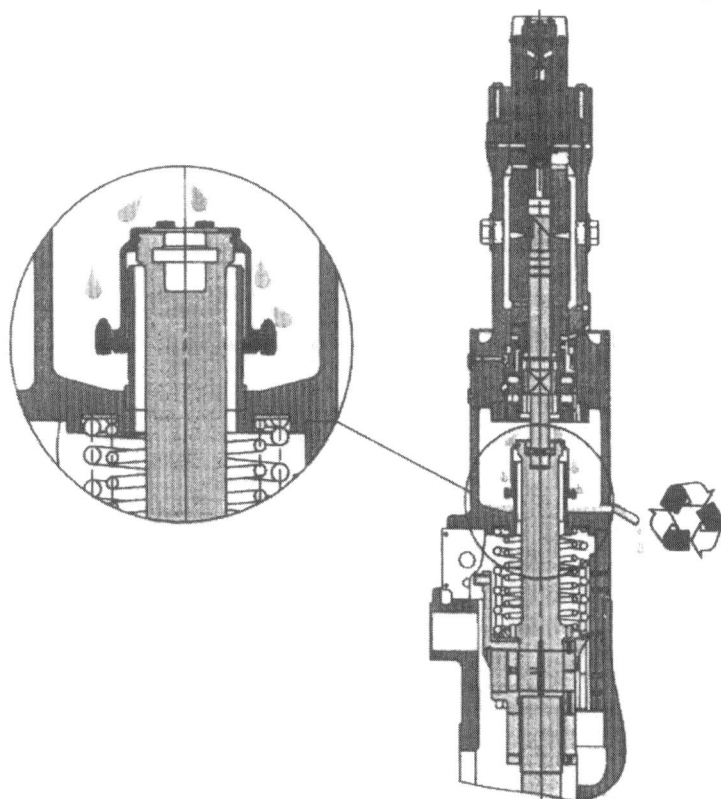

Figure 56. MAN-B&W Fuel Pump with 'Umbrella' Sealing

Advantages claimed for this are as follows:

- No separate camshaft lubricating oil system required.
- Lower installation costs.
- Lower maintenance costs.
- Clean drain fuel oil for recycling.

Leaking fuel pump relief valves may cause a reduction in the exhaust temperature of the affected unit. The leakage pipe will also be hot if the engine is operating on heavy fuel oil. Figure 57 shows injection pressure curves and how faults affect them.

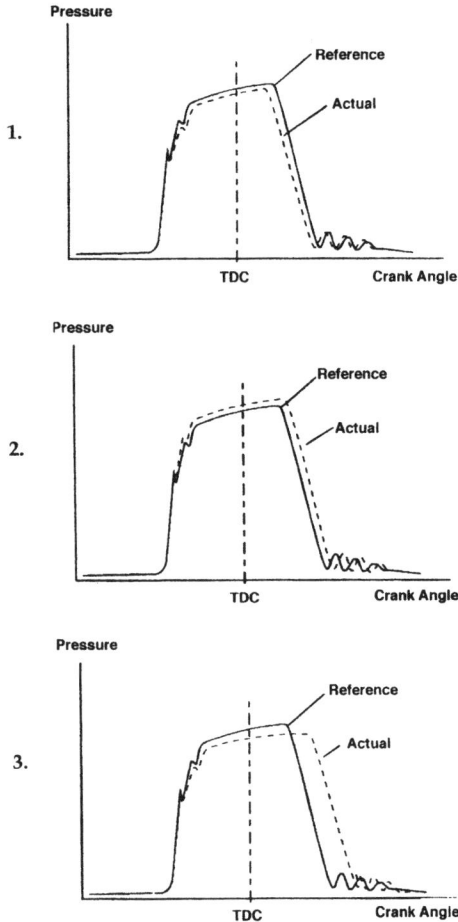

Figure 57. Fuel Injection Curves showing Faults

Curve 1   Maximum injection pressure reduced. Shortened injection period. Possible causes: worn or damaged injector nozzle; low fuel viscosity (all cylinders affected).

Curve 2   Maximum injection pressure increased. Increased injection period. Possible causes: choked injector nozzle; high fuel viscosity (all cylinders affected).

Curve 3   Maximum injection pressure reduced. Increased injection period. Possible causes: worn or damaged fuel pump; leaking suction valve.

Fuel Pump Maintenance

As with an injector needle and guide, the fuel pump plunger and barrel are ground as a matched pair and should not be interchanged. If a problem occurs with a pump, or it is suspected as faulty, it is often easier to change the complete pump and overhaul the defective unit later. After changing a pump the injection timing should be checked. Spare plungers and barrels are usually supplied coated in wax, which should not be disturbed until the parts are ready for use.

When overhauling fuel pumps utmost cleanliness should be observed. Parts should be thoroughly washed in kerosene and blown clean with dry air. All seals and O-rings should be renewed and all ground surfaces and seats checked for damage. If special grinding in tools are not available on board the defective components should be sent to a specialist reconditioner. Plungers which have seized in their barrels should be discarded.

Whenever a new plunger and barrel have been fitted, a new pump fitted or the existing pump removed and replaced, the fuel pump timing should be checked. This may also need to be adjusted if indicator cards show early or late injection on a cylinder (see Section 2.2.8).

Traditionally fuel pump timing was carried out by checking the flywheel position when the ports in the barrel were just covered by the plunger. This was achieved by removing the pump non return valve and turning the engine until fuel just stopped pouring from the delivery pipe. Another method included removing plugs in the barrel and shining a light through the ports. The engine was then turned until the light was not visible. The crank angle was noted and checked against the manufacturer's value.

More modern and accurate methods involve inserting measuring devices above the barrel to measure the position of the barrel when the engine piston is at top dead centre.

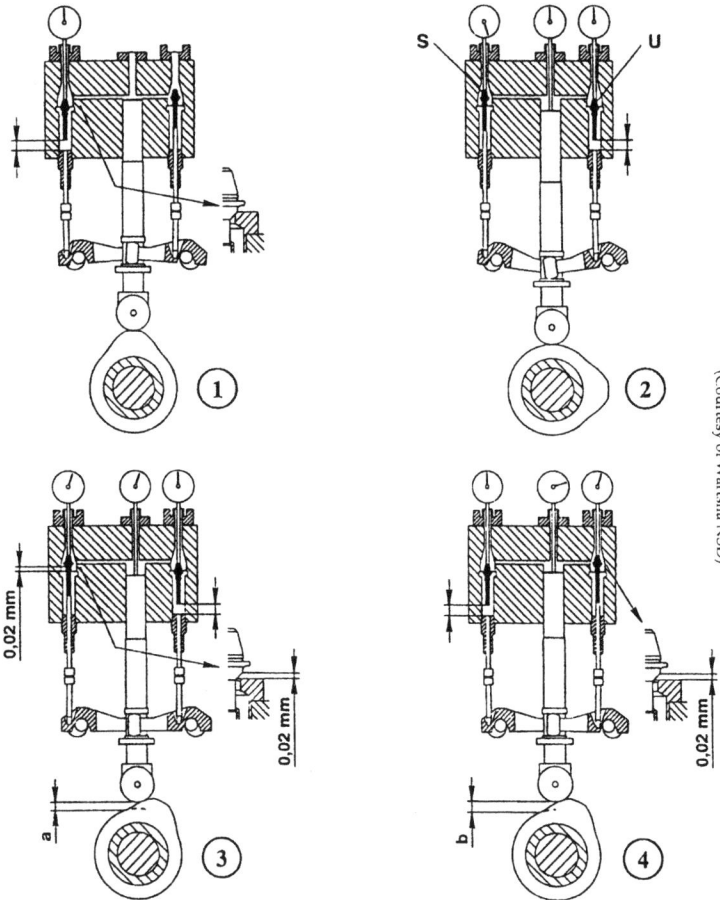

Figure 58. Wärtsilä NSD RTA Fuel Pump Timing

Procedure for Wärtsilä NSD RTA Fuel Pump Timing

1. The engine should be turned in the ahead direction until the pump plunger is at top dead centre. The dial gauge should be fitted to the suction valve (S). After tensioning the gauge it should then be set to zero.

2. The engine should be turned in the astern direction until the fuel pump roller is on the base of the cam, as shown. Dial gauges should be fitted to the plunger and spill valve (U). After pre-loading the gauges they should both be set to zero.

3. The engine should be turned in the ahead direction until the suction valve gauge reads 0.02mm. At this point the suction valve is just closing and fuel delivery will begin. Note the travel of the plunger gauge (a) and the crank angle.

4. The engine should continue to be turned in the ahead direction until the spill valve gauge reads 0.02mm. This indicates the spill valve is just opening and fuel delivery will stop. Again note the crank angle and the plunger stroke (b). The effective plunger stroke = b-a. Upon completion all values should be checked against the engine manufacturer's values.

The fuel pump linkage should be regularly inspected and lubricated. Engine vibration can result in linkages becoming loose and pins and bushes can wear causing excessive play in the linkage. This can induce high peak pressures and lead to fluctuating engine speed.

2.2.7   Indicator Diagrams

Indicator cards are used to determine compression pressure, peak pressure and engine power. They should be taken on a regular basis as they are vital for assessing whether any adjustments to the fuel timing are required. They also determine the condition of the cylinders.

Traditionally, indicator cards were taken using a mechanical indicator fitted to the cylinder indicator cock. This method is still in use today, however, more and more vessels are now equipped with electronic engine performance monitors. These monitors, as well as having sensors fitted to the indicator cocks, also monitor shaft rpm, scavenge air pressure and fuel pump pressure (discussed in Section 2.7).

High cylinder pressure leads to an increase in mechanical stress on the running gear and may result in fatigue cracking. Loads on bearings are also increased, reducing oil film thicknesses, and problems may occur (particularly on the crosshead bearing).

If the cylinder pressure is excessive then the cylinder relief valve should lift. It is important that these valves are regularly tested to ensure they lift at the correct pressure.

There are essentially two types of indicator card - a power card and a draw card (out of phase card). These are shown in Figure 59.

The draw card is perhaps the most useful type of indicator card. It is used for setting up the engine fuel pump timing and provides more information about the cylinder condition than a power card. Furthermore, not all engines can take power cards as the indicator drum needs to be driven by the engine.

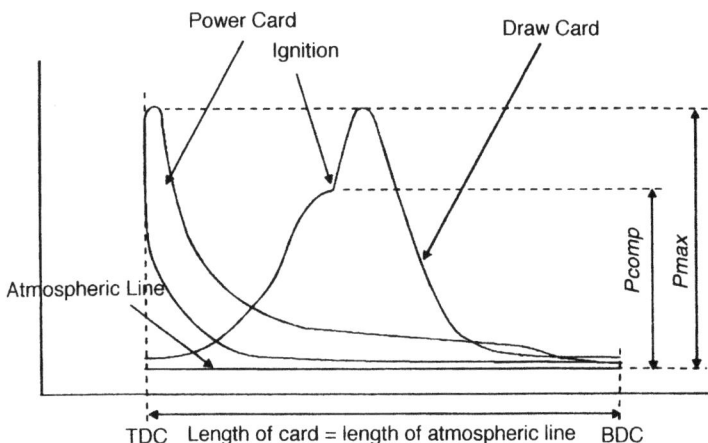

Figure 59. Indicator Diagrams

Prior to taking cards the indicator cock should be blown through to ensure it is clear. Fit the indicator to the indicator cock and fit the card to the drum. Check the stylus is not pulsating, due to leaking indicator cock packing. Press the stylus onto the card and pull the drum cord to draw the atmospheric line.

For the power card, connect the drum cord to the engine indicator linkage and the drum will then start oscillating. The indicator cock can then be opened and the stylus pressed onto the card to draw the power card.

For the draw card, the indicator cord must be disconnected from the engine drive and the drum oscillated manually by pulling the cord. The stylus should be pressed onto the card just before the stylus rises. This quite often takes some experience to get the timing of the two operations correct.

<u>Interpretation of Indicator Cards</u>

Normal Diagram

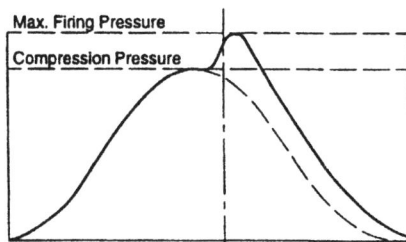

P_{COMP} Normal, P_{MAX} High
(Ignition Advanced - Too Early)

- VIT incorrect
- Fuel Pump Lead too great
- Adjust FQS if on all cylinders (Sulzer)

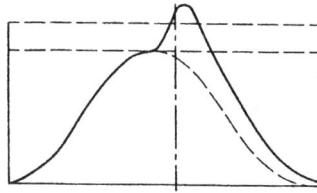

P_{COMP} Normal, P_{MAX} Low
(Ignition Retarded - Too Late)

- Poor Fuel Quality (If on all Cylinders)
  Correct with FQS (Sulzer)
- Fuel Pump Suction Valve Defective
- Nozzle Holes Blocked or Fouled
- Defective Fuel Injector
- Fuel Pump Lead too Small
- Fuel Pressure too Low

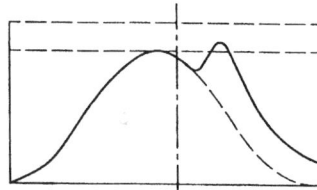

P_{COMP} Low, P_{MAX} Low

- Leaking Exhaust Valve
- Charge Air Pressure too Low
- Burnt Piston Crown
- Worn or Broken Rings
- VEC Timing Incorrect (Sulzer)

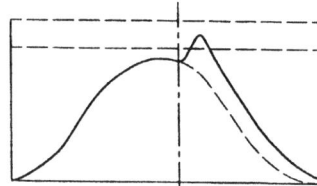

P_{COMP} High, P_{MAX} High

- Engine Overload
- VEC Timing Incorrect (Sulzer)

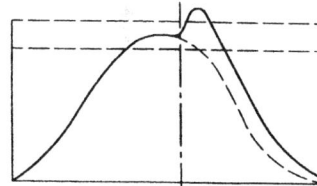

(Courtesy of Wärtsilä NSD)

Figure 60. Interpretation of draw cards

When the indicator cards are taken the following data should be recorded:

- Engine rpm.
- Load indicator.
- Cylinder exhaust temperatures.
- Scavenge air temperatures.
- Scavenge air pressure.
- Air cooler inlet & outlet air temperatures.

- Air cooler differential pressure.

- Turbocharger rpm and temperatures.

- Fuel temperature and any other fuel analysis data.

- Cylinder running hours - in total and since last overhaul.

- Injector hours since last changed or overhauled.

- Ship speed and slip.

- Draught forward and aft.

- Fuel consumption.

- Cylinder oil consumption.

- Ambient air temperature.

All the above should be kept on a record sheet, together with the compression and peak pressures. These can then be referred to should problems arise.

Compression and peak pressures should all be fairly even, within 2 or 3 bar of each other. Any significant differences should be investigated to determine the cause. A reduction in compression pressure may indicate wear to the rings, liner, piston or valves on the affected unit. Any significant difference in the peak pressures may be due to incorrect timing on the affected unit(s).

Calculation of Engine Power

If the engine is equipped to take power cards, these need to be measured after they have been taken to determine the area of the diagram. This is then used to calculate the indicated power. A planimeter is used for this purpose and is shown in Figure 61.

Select a starting point on the curve and place the tracer centre circle over the point. Using the rollers, set the planimeter counter and vernier to zero and then start tracing around the diagram, keeping the line within the tracer circle. Take an average reading by tracing around the diagram three times and dividing the final reading by three. This gives the area (a) of the diagram in mm^2.

Measure the length of the atmospheric line. This is the length of the indicator diagram, l (mm). The Mean Indicated Pressure (p) is then calculated as follows:

$$p = \frac{a}{l \times C} \ N/m^2$$

Where C = Indicator spring rate (mm per N/m^2)

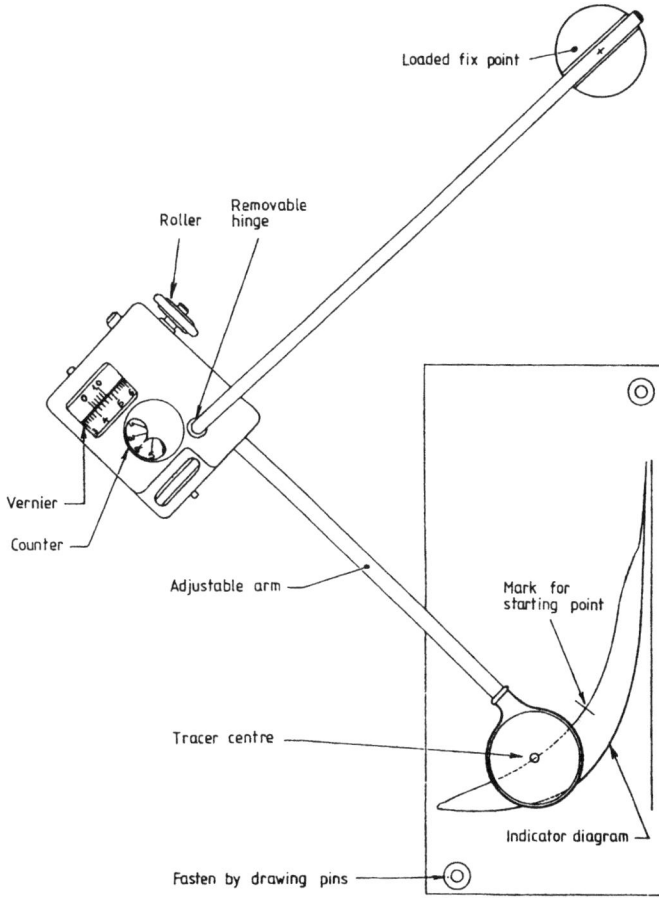

(Courtesy of MAN-B&W Diesel)

Figure 61. Planimeter for measuring area of indicator diagram

The mean indicated pressure is then used to determine the indicated power as follows:

$$\text{Indicated Power (pi)} \quad = \frac{pLAn}{60} \text{ (Watts)}$$

Where:
P = Mean Indicated Pressure (N/m^2)
L = Length of Piston Stroke (m)
A = Area of Cylinder (m^2)
n = Revs per Minute

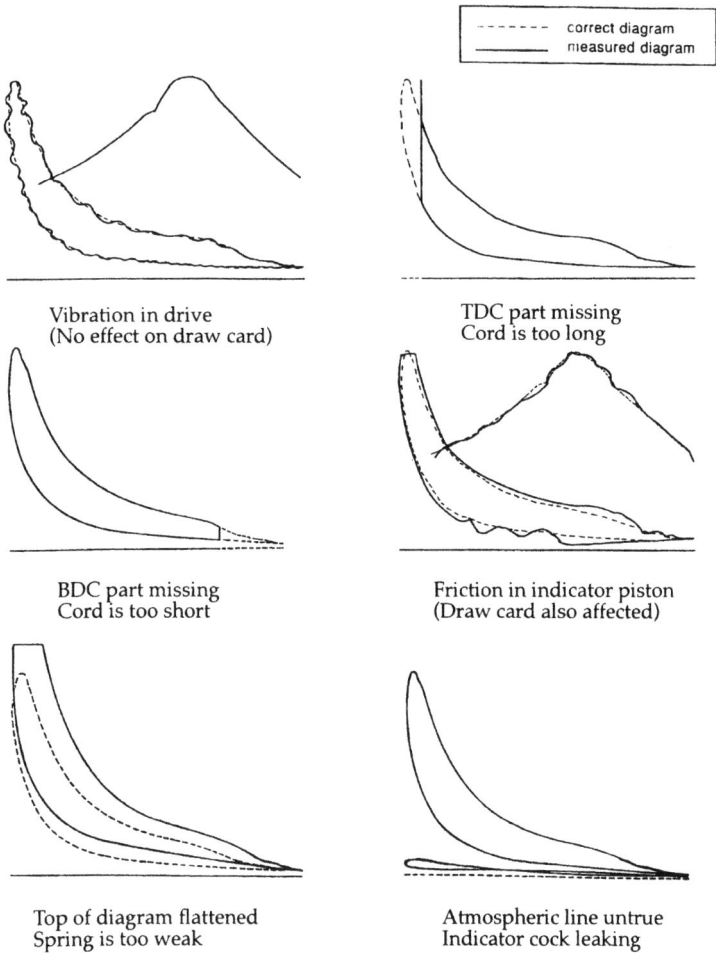

Vibration in drive (No effect on draw card)	TDC part missing Cord is too long
BDC part missing Cord is too short	Friction in indicator piston (Draw card also affected)
Top of diagram flattened Spring is too weak	Atmospheric line untrue Indicator cock leaking

(Courtesy of MAN-B&W Diesel)

Figure 62. Faulty indicator cards

Care of the Indicator

- Take care not to damage the linkage as this will impair accuracy.
- Do not allow the indicator to overheat. Keep the indicator cock open for the shortest possible time.
- Always dismantle, clean and lightly oil the indicator after use.
- Always keep it in its box.

Balancing Medium Speed Engines

Due to the speed of the engine it is not possible to take indicator diagrams using a conventional mechanical indicator. In this case the maximum cylinder pressures are recorded using a peak pressure indicator. When using a peak pressure indicator to check engine timing knowledge of the state of individual cylinder components is essential. If these are known to be in good condition then peak pressures can be taken as a guide to injection timing.

2.2.8   Bearings

Bearing Construction

Bearings are fitted in order to reduce both friction and wear between the running surfaces. This is achieved primarily by the lubricant, with a hydrodynamic oil film to keep the surfaces apart. The oil film can never completely prevent contact, particularly at times such as start-up. Dirt particles larger than the film thickness may also enter the bearing and damage the journal, therefore, a suitable bearing material must be used. Bearings require the following properties.

Compatibility   Bearing and journal materials must be compatible to reduce any tendency to local welding between the bearing alloy and any surface asperities on the bearing journal.

Conformability   The bearing lining will 'conform' to any slight misalignment.

Embeddability   The material must be soft enough to embed hard particles that may otherwise damage the journal.

The latter two properties require the bearing alloy to be soft, however, it must also be strong. The material chosen must be a compromise between the two properties to be acceptable for the intended bearing duty.

Engine bearings are usually of a thin or thick shell type. The difference is in the thickness of the steel backing. Thin and thick shell bearings are shown in Figure 63.

Tangential Runout

Tangential runout gives a gradual run in for the oil and prevents any oil scraping effect. It also reduces the resistance to flow.

Bore Relief

Bore relief will compensate for any slight misalignment. If misalignment occurs there will be a slight protruding edge that will act as an oil scraper which may starve the bearing surface of oil.

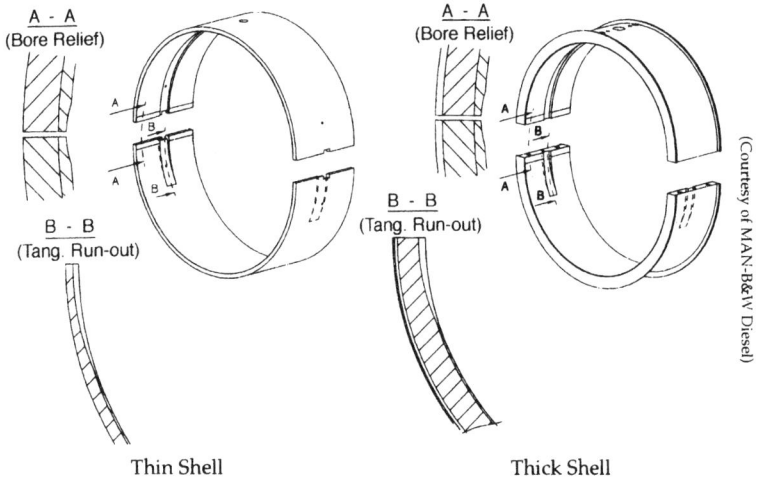

Thin Shell   Thick Shell

Figure 63. Thin and thick shell bearings

Oil Wedges & Grooves

Oil wedges and grooves provide a better oil distribution over the bearing surface, improving lubrication and cooling. The oil wedges assist in the formation of a hydrodynamic oil film.

White Metal Bearings

Shell bearings consist of a steel backing with a white metal lining bonded to it. The steel backing gives support to the bearing and improves fatigue life.

White metal bearings may be tin based (typically 89 per cent tin, 7.5 per cent antimony and 3.5 per cent copper) or lead based (typically 83 per cent lead, 15 per cent antimony, 1 per cent tin and 1 per cent arsenic). Other trace alloying elements are usually added to improve the grain structure. Tin based white metals are more commonly used as they have better fatigue strength and corrosion resistance.

There may be a bearing overlayer consisting of a galvanic coating which is typically 90 per cent lead and 10 per cent tin. This ensures good embeddability and conformity between surfaces. In addition there may be a flash layer of 100 per cent tin to prevent corrosion. This layer of tin is then removed during the running in process.

Aluminium Tin Bearings

In many medium speed engines aluminium tin bearings are used (AlSn20), which contain 20 per cent tin. Aluminium tin has a higher fatigue strength than white metal, but also possesses good conformity and compatibility. A thin galvanic layer may be applied over the bearing surface to improve corrosion resistance.

Crosshead engines also use bearings with 40 per cent tin (AlSn40), which has similar surface properties to white metal with better fatigue resistance. A multi-layer aluminium tin bearing as used on a medium speed engine is shown in Figure 64.

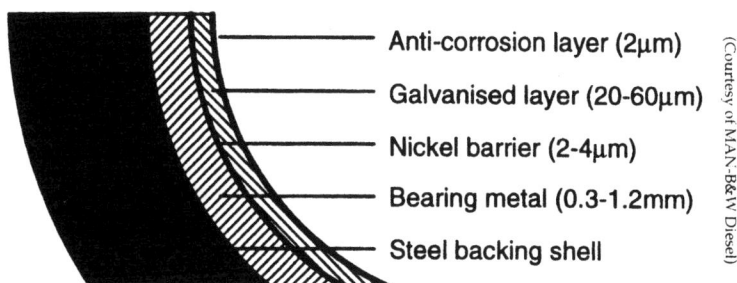

- Anti-corrosion layer (2μm)
- Galvanised layer (20-60μm)
- Nickel barrier (2-4μm)
- Bearing metal (0.3-1.2mm)
- Steel backing shell

(Courtesy of MAN-B&W Diesel)

Figure 64. Multi-layer shell bearing

Bearing Maintenance & Problems

Bearings bed themselves in over a long period of time and should be removed as little as possible. Bearings should be maintained regularly and every six months the clearances should be measured. Clearances are usually measured with feeler gauges, however, on main bearings, where the bearing cap prevents access with normal feeler gauges, 'Swedish' feeler gauges are used. They are enclosed in a long tube and can reach down to the journal. (See Figure 65.)

Bearing wear on large engines is usually in the order of 0.01mm/10 000 hours, however, the bearing clearance is not a true indicator of bearing wear. Wear of bimetal shell bearings is determined by measuring the wall thickness and comparing with the original thickness. Bearings that have a galvanised layer should be replaced when a predefined percentage of the layer has worn away, as stated by the engine manufacturer.

Bearing shells are usually turned out with special tools, determined by the size of the bearing. Care should always be taken when turning the engine, to prevent damage or distortion to the bearing shell. It is important when turning out a bearing shell that the shell is continually checked to ensure it is actually turning with the shaft.

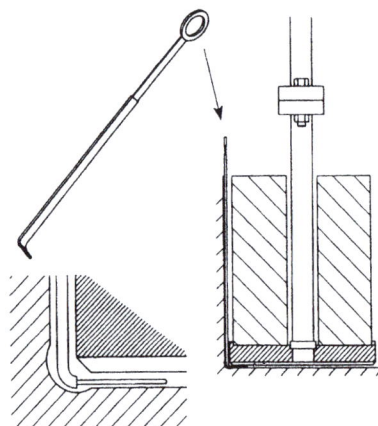

(Courtesy of MAN-B&W Diesel)

Figure 65. Measuring main bearing clearance

If a bearing has to be opened for inspection the clearance should be measured, both before dismantling and then again after reassembly. Once removed, the bearing shells should be inspected for signs of cracking, scoring or wiping. The backs of the shells and the bearing pockets should then be checked for signs of fretting. The journal should also be examined for any scoring or cracks, particularly around oil holes and in the roots of fillets.

Bearing bolts should be hung up and tapped with a hammer. This should produce a ringing sound - a dull sound may suggest a defect within the bolt. Bearings should be left covered if opened for any length of time, such as when waiting for a surveyor, to prevent damage and dirt ingress.

When refitting the bearing the back of the shell and the bearing housing should be thoroughly cleaned. The journal, bearing surface and bearing back should be lubricated prior to fitting. The bearing tongues should be fully seated in their recesses.

The bearing running surface should normally be a matt grey colour for white metal and aluminium tin bearings.

Wiping

Wiping occurs when the white metal has been overheated, causing it to melt and be carried along with the journal producing a smeared effect. At cooler parts of the bearing the metal will solidify. Any high spots will be polished off by the journal to give a shiny appearance.

If metal to metal contact occurs the high temperature generated will soon melt the white metal. Metal to metal contact will occur due to a breakdown of the oil film caused by either overloading or an insufficient or incorrect lubricant. Insufficient pre-lubrication is the usual cause of wiping to the galvanic layer.

Scoring

Scoring is visibly evident as lines of varying depth and number in the direction of rotation. The severity depends upon the cause of the problem and may be one or two light scores or many deep ones that expose the backing. Scoring is caused by hard particles that are larger than the oil film thickness. These may embed themselves in the bearing and can damage the journal, particularly at start up or when slow running. The presence of these particles suggest inadequate filtration and/or centrifuging.

Light scores are satisfactory if they cannot be measured or felt with a finger. If the scores are severe, if they are wide or can be measured, the shell should be replaced. If not severe the bearing may be left in use but any sharp edges to the scores should be removed.

Crazing

Crazing occurs due to the formation of fatigue cracks caused by overloading the bearing. These small cracks will join into a network of cracks to give a crazed effect to the surface. This means the bonding has failed and small pieces of white metal have become dislodged, exposing the bearing backing. Small cracks on the bearing surface are satisfactory, but the white metal bond should be tested by lightly tapping the white metal with a wooden hammer handle. Once a surface has become crazed the shell should be renewed.

2.2.9   Crankshaft Deflections and Crankcase Inspection

Crankshaft Deflections

Crankshaft deflections are taken to ensure the axis of the crank journals has not deviated from the theoretical axis. Deflections are normally taken once per year, however, they should always be taken again when any bearings have been renewed (on older engines deflections should be taken every six months). Deflections should also be taken if a vessel has grounded or there has been any change to the chocking arrangement.

Deflections should always be taken while the vessel is afloat. The crankcase temperature and draught, forward and aft, should also be noted. If possible, deflections should be taken within similar temperature and trim to previous readings to ensure a good comparison can be made. Indicator cocks should always be opened as the deflections will be affected if these are left closed.

Deflections are taken by fitting a dial gauge indicator between the crank webs and setting it to zero. Usually centre punched marks are present to locate the gauge.

The normal sign convention is that a positive reading shows the webs have opened out and a negative reading shows the webs have closed in.

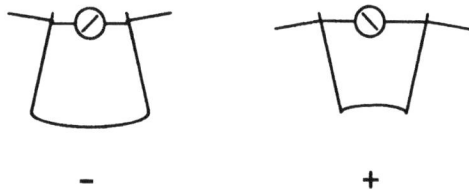

Figure 66. Deflection Gauge Sign Convention

The crankpin should be as near to bottom dead centre as possible. The shaft is then turned with the turning gear, or slowly barred over. Readings are then taken at the points shown in Figure 67.

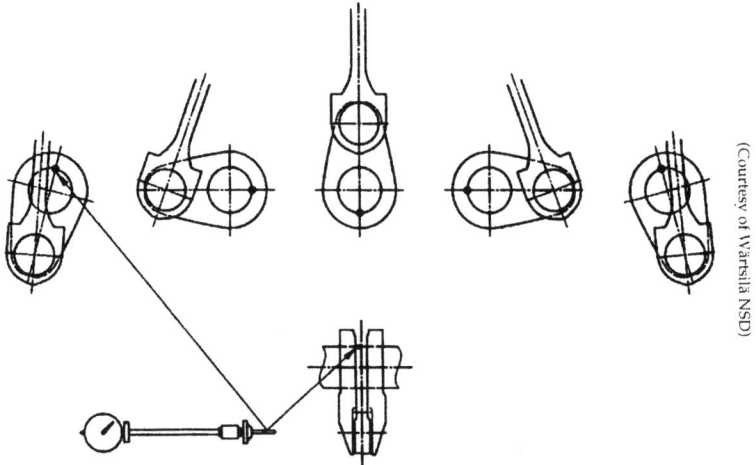

(Courtesy of Wärtsilä NSD)

Figure 67. Crankshaft Deflections

Engine manufacturers usually have limits for the difference between the top and bottom dead centre readings. The final reading should return to zero. If not, an average of the start and finish readings should be used instead.

Even though the alignment may be correct there will still be a deflection reading due to the weight of the running gear. It is the magnitude of the deflection that is important, however, and how often the sign changes. The following factors can influence the recorded readings.

- Bearing weardown.
- Deflection of adjacent cranks.

- Weight of running gear.
- Tension of camshaft chain drive.
- Incorrectly tensioned tie bolts.
- Thermal effects, e.g. warm engine room and cold sea water.
- Wear of chocks.
- Alignment of propeller shafting / generator.
- Vessels trim and condition, i.e. ballast or loaded.

Crankcase Inspection

This should be carried out at regular intervals for both slow and medium speed engines. On larger engines it is possible to enter the crankcase to carry out an inspection. An inspection mirror can be used on smaller engines, however, make sure it is attached to a length of cord in case it drops into the sump tank.

The engine must be allowed to cool down for 30 minutes before opening any crankcase doors. Once the doors have been opened the crankcase should be ventilated. Before entering the crankcase the turning gear must be engaged, the starting air valve closed and the air line drain open. Indicator cocks must also be open.

Each unit should be inspected in turn, working from top to bottom or vice versa. The bottom of the crankcase should be checked for any debris, rags, metallic particles, etc., particularly around the drain into the sump tank. The presence of any water and/or build up of sludge deposits should also be checked, particularly on ledges and in cooler parts of the crankcase.

All locking devices such as locking wire, split pins, tab washers, etc., should be checked for security. All bolts should be visually examined and checked by hand to ensure none are loose. Any signs of cracking in the bedplate, in way of the main bearing seat, and also the A frame, in way of the crosshead guides, should also be checked.

2.2.10   Crankcase Explosions

For an explosion to occur there needs to be an air/oil mixture within the explosive limit and an ignition source. Under normal conditions the crankcase is full of air so the mixture is well outside this limit.

The oily atmosphere inside the crankcase is usually produced by large oil droplets caused by the splashing oil. If a problem occurs causing an increase in temperature, such as a bearing running hot or piston rubbing on the liner, the oil will become vapourised and start to form an oil mist. Gradually the mixture in the crankcase will get richer in oil and fall within the explosive limit. All that is now required is an ignition source. This could be the piston blowby or the hot spot itself.

By the time ignition occurs the mixture may be very rich and will therefore be slow burning. Even so, the ignition will cause a rise in pressure which may be sufficient to blow off crankcase doors. Once this occurs the pressure in the crankcase will fall rapidly. The inrush of air through the damaged crankcase will result in a much more severe secondary explosion. A plot of pressure against time is shown in Figure 68.

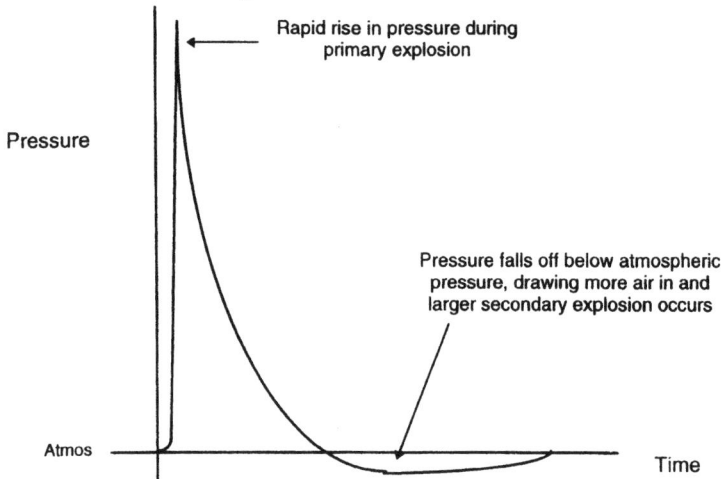

Figure 68. Pressure Rise due to Crankcase Explosion

Crankcase relief devices are fitted to prevent damage to the crankcase and an inrush of air. These spring loaded devices lift to relieve the pressure and then re-seat to prevent any air entering the crankcase. A typical crankcase relief device is shown in Figure 69. Classification Societies specify the minimum opening sizes for these based upon the volume of the crankcase. The number of devices required depends on the size of the cylinder bore.

Oil Mist Detectors

To detect an overheating bearing, temperature sensors would have to be fitted to each bearing. This is not a practical proposition, therefore, an oil mist detector can be fitted to detect the developing oil mist before it reaches sufficient concentration to cause an explosion. The detector consists of a fan which draws air samples from each crankcase section in turn and passes the sample through a detector cell. This cell uses a light scatter technique to measure the density of the oil mist. Once the mist density reaches a predetermined level, which is much lower than the lower explosive limit of 50mg/l, an alarm will sound and there will usually be an automatic engine slow down.

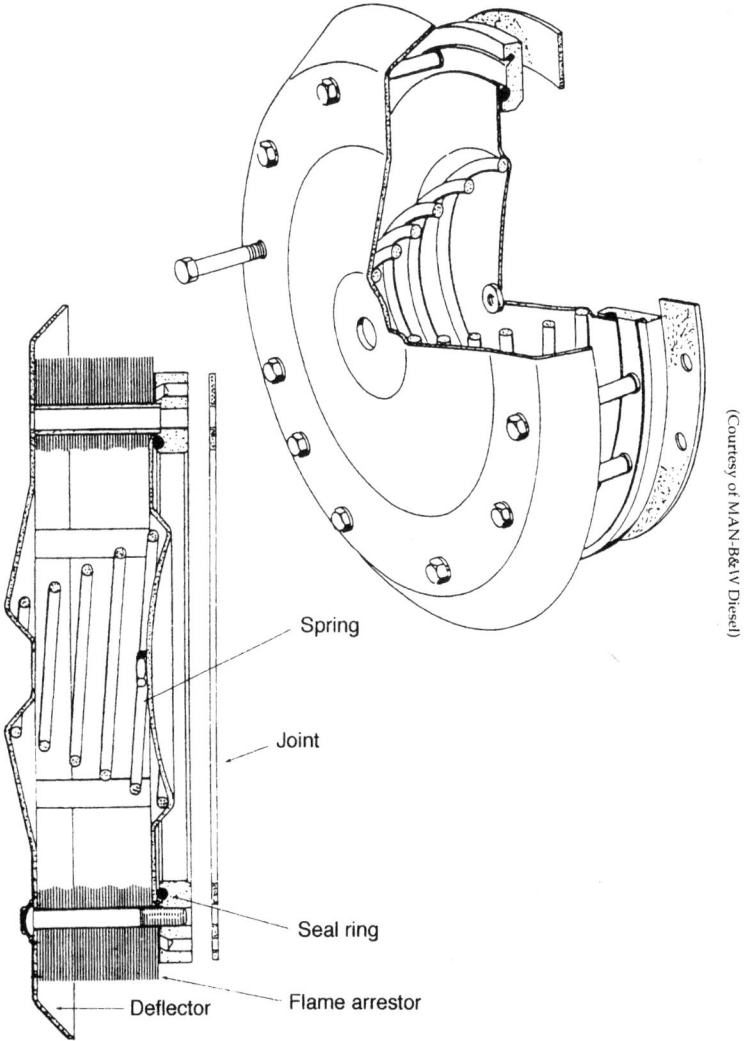

(Courtesy of MAN–B&W Diesel)

Spring

Joint

Seal ring

Deflector   Flame arrestor

Figure 69. Crankcase Relief Device

Maintenance of the oil mist detector mainly consists of cleaning the lens and fan filters. Some systems have alarms to warn when these need cleaning. Cleaning should be carried out fairly regularly if false alarms are to be avoided. The lens will eventually become coated with oil and condensation which should be cleaned off with soapy water and a soft cloth. Solvents should not be used. The detector head and/or fan filters should be cleaned in the same way. The oil mist detector should be inspected daily to ensure it is going through its normal operating sequence. The following procedure should be carried out in the event of oil mist formation.

- Reduce engine speed - this may be automatic.
- Check with the Bridge that it is safe to stop the engine.
- Stop the engine and engage the turning gear.
- Stop the fuel pumps and auxiliary blower(s).
- Vacate the engine room and allow at least 20 minutes before re-entering.
- Stop main lubricating oil (LO) pumps and open the crankcase doors.
- Inspect all bearings and running surfaces for any hot spots, including the thrust bearing.
- Inspect the bottom of the crankcase for any signs of bearing metal.
- If evidence of overheating is found the affected bearing must be opened up for inspection.
- If the cause of the oil mist cannot be found it may be due to a fault with the oil mist detector, but this should not be an automatic assumption. Other causes may be a piston or stuffing box blowby (check crankcase pressure) or excessive water in the lubricating oil. Check that all engine pressures and temperatures were normal before the oil mist alarm occurred.
- Start the main LO pump and turn the engine with the turning gear. Check oil is flowing freely from all the bearings and gear/chain sprayers.
- Close the crankcase doors and start the engine.
- Stop the engine after running at slow speed for 15 minutes. Check all the bearings as before. Repeat after running for 15 minutes at full load.

2.2.11   Bedplates and Frames

Figure 70 shows some of the forces acting on a crosshead engine structure.

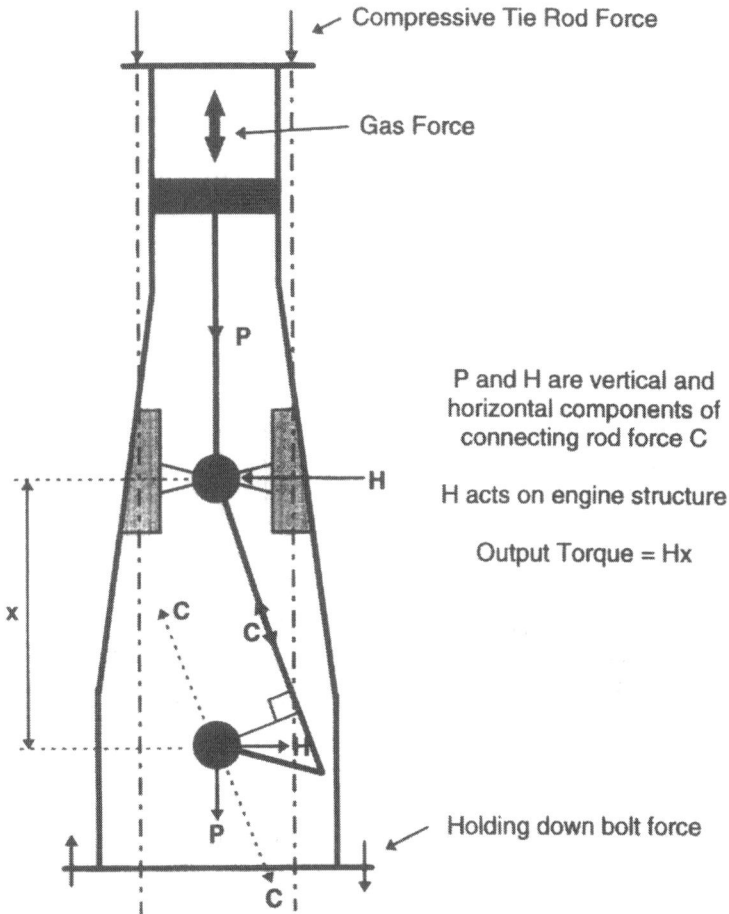

Compressive Tie Rod Force

Gas Force

P

P and H are vertical and
horizontal components of
connecting rod force C

H acts on engine structure

Output Torque = Hx

H

x

C

C

H

P

C

Holding down bolt force

Figure 70. Forces on Engine Structure

The engine bedplate maintains the longitudinal, transverse and vertical alignment of the engine. On crosshead engines the bedplate, A frames and cylinder jackets are all held together by tie bolts which are tensioned so as to maintain the assembly in compression at all times.

Large engine bedplates are usually fully welded and can suffer from cracking, particularly where the bearing seat is welded to the transverse girders and the transverse girders join the longitudinal box sections.

The traditional construction was a fabricated box section transverse girder and welded in bearing seat. The welded in seat had a low fatigue strength which eventually resulted in cracks forming. This was overcome by using a cast bearing saddle to reduce the severity of the change of section. This also made the fatigue strength much higher.

Figures 71 and 72 show typical bedplate construction for both earlier and later designs of Wärtsilä NSD crosshead engines. The tie rods are located as near as possible to the shaft centre line to minimise bending of the transverse girder.

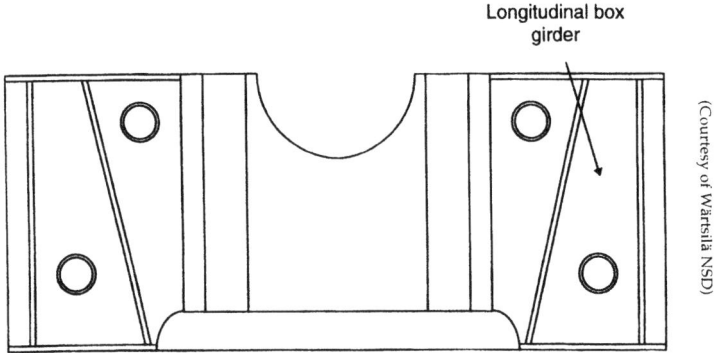

Figure 71. Bedplate Construction for Wärtsilä NSD RND Engine

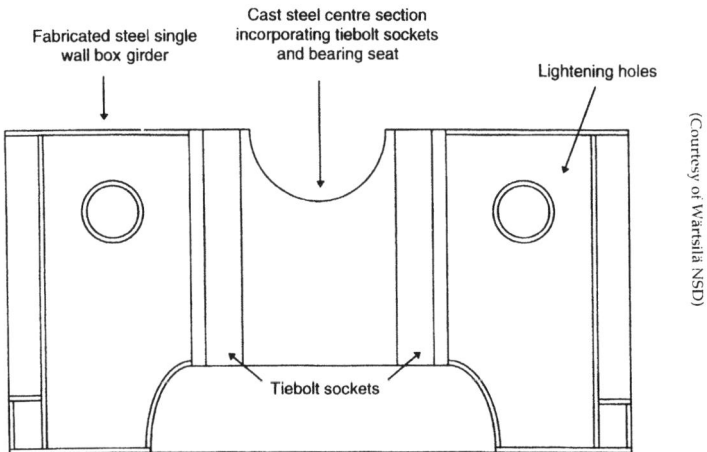

Figure 72. Bedplate Construction for Wärtsilä NSD RTA 84 Engine

The RTA design has changed slightly and has a single wall box girder. The crankshaft is semi-built on the larger models, however, it may be a solid forging on small bore crosshead engines.

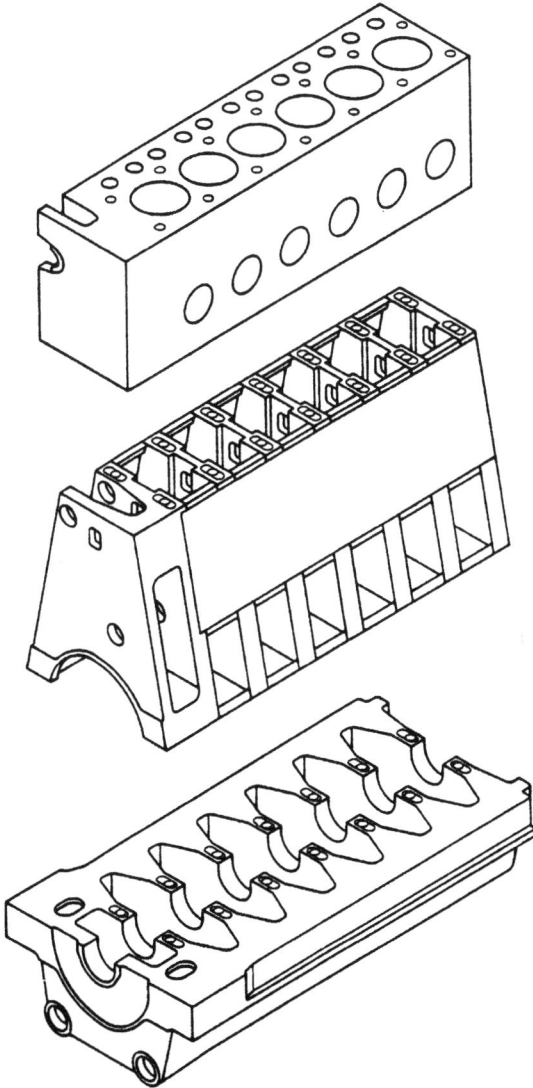

(Courtesy of Wärtsilä NSD)

Figure 73. Bedplate for Wärtsilä NSD RTA 38

Some of the smaller crosshead engines have a cast bedplate (shown in Figure 73). Most medium speed trunk piston engines will have a cast crankcase and a cast or fabricated sump with a solid forged crankshaft. Two typical examples are shown in Figure 74.

(Courtesy of MAN-B&W Diesel)   (Courtesy of Wartsila NSD)

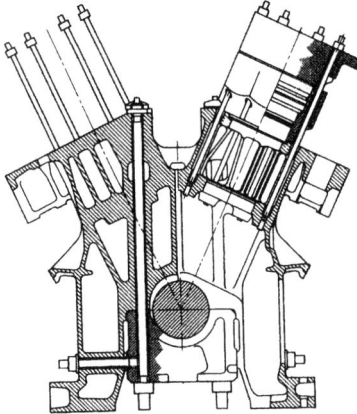

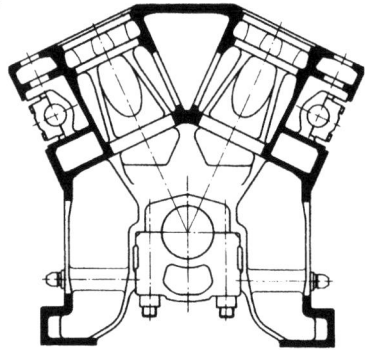

MAN-B&W V48/60   **Wärtsilä NSD ZAV40**

Figure 74. Trunk Piston Engine Crankcase

Checking Tension of Tie Bolts

Tie bolts keep the engine assembly (bedplate, A-frame and jacket) together and are pre-tensioned to keep the whole assembly in compression at all times (see Figure 75).

The bolts prevent frame feet separation under firing loads which would lead to fretting. As the bolts are extremely long they have a tendency to vibrate, therefore, supports are fitted to reduce the vibration amplitude.

Tie bolts are usually hydraulically tensioned and should be checked annually for tightness. Loose bolts may fret which can usually be identified by the presence of red oxide dust on the landing face next to the bolt. This usually mixes with oil present on the top of the jackets to form a red coloured paste. If fretting is allowed to continue it may eventually wear the jacket landing face and the underside of the tie bolt nut. Tightening may then increase the stress in the bolt and possibly shorten the bolt life.

Before fitting the hydraulic nuts the area around the tie bolt nuts, and the threads, should be thoroughly cleaned. Bolts should have covers to protect the threads. Two opposite bolts should be checked at the same time.

Fit the hydraulic nuts and connect the hose from the nuts to the hydraulic pump. Tighten the hydraulic nuts right down and then slacken back half a turn. Increase the pressure with the hydraulic pump to reach the normal tightening pressure.

Check the tie bolt nuts are tight and use a feeler gauge to check the bolts are firmly down on the landing face. If tie bolt threads have been damaged the nut may be tight, even though it is not seating on the landing face.

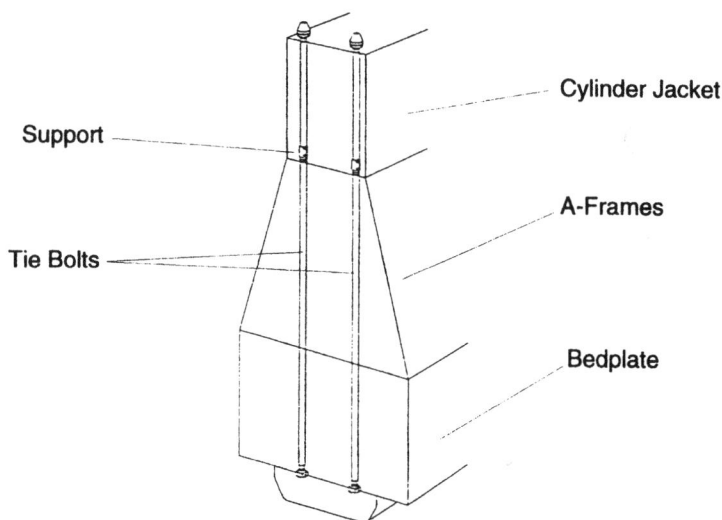

Figure 75. Tie Bolts Assembly

Holding Down Bolts

The main purpose of the holding down bolts is to restrain the reaction torque from the propeller. The engine bedplate sits on machined chocks which sit on the foundation plate. The holding down bolt passes through the chock as shown in Figure 76. Side and end chocks restrain the engine transversely and axially respectively.

As with tie bolts, holding down bolts should be checked annually for any loose bolts, as fretting can also be a problem. Fretting can lead to wear of the landing faces of bedplate, chock and foundation plate. As they are located under the engine room floor plates their condition may not be regularly checked. They will rust if not protected, therefore, they should be coated with a protective grease or lubricant.

Bedplate

Cast iron chock

Foundation plate

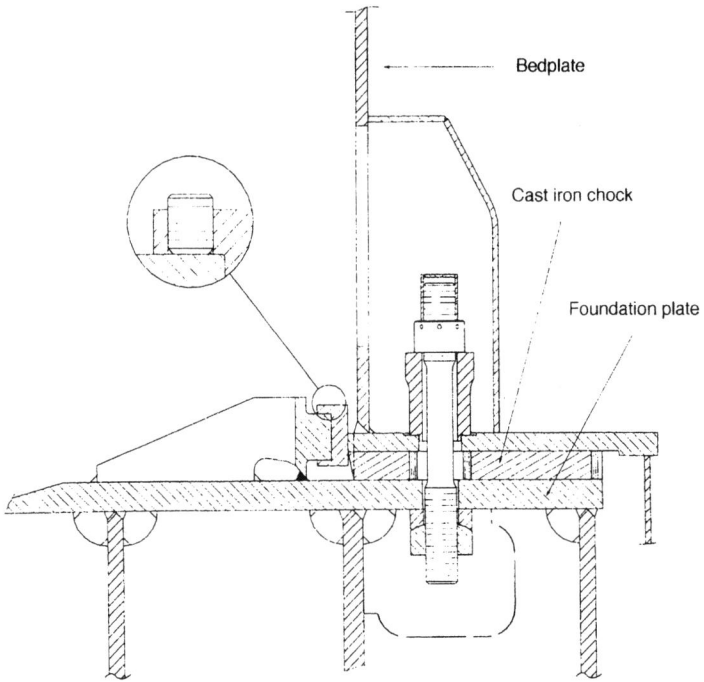

(Courtesy of MAN-B&W Diesel)

Holding Down Arrangement

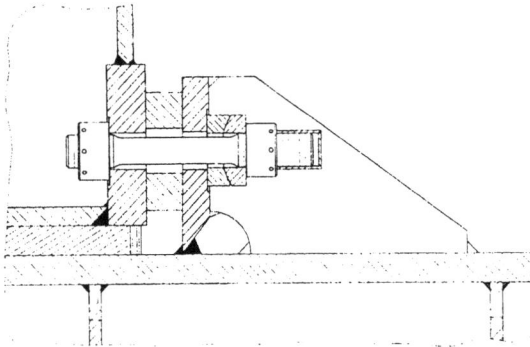

Side and End Chocking

Figure 76. Holding Down Bolts and Side Chocks

Bolts on larger engines are usually hydraulically tightened so checking tightness can be a laborious job. On smaller engines a torque wrench may be used. If checking hydraulically, the bolts should be gradually tensioned up to the normal pressure and the pressure noted where the bolt can be tightened. The bolt is then tightened at the normal pressure. A record of the pressures should be kept. Do not exceed the normal tightening pressure.

A bright lamp should be used to thoroughly check the area around the bolt lands for any signs of fretting. Any rusted bolts should be cleaned with a wire brush and coated with protective grease.

Non Metallic Chocks

Many engines are now mounted on resin chocks. First the engine is aligned with the jacking bolts then resin is poured in. The resin is a two pack liquid which has to be mixed just prior to pouring. A dam must be fitted around the area being chocked and the bedplate and foundation must be thoroughly cleaned and de-greased. Resin chocks have the following advantages:

- Cheaper.
- Eliminate fretting.
- Reduced maintenance.
- Increased reliability.
- 100 per cent contact over irregular surfaces.

2.2.12   Gears and Gearboxes

Vessels with medium speed engines driving a propeller require a gearbox to give the required propeller speed. The gearbox can also have an output shaft to drive a shaft generator.

The gearbox will have its own lubricating oil circulation system with pump, filters and a cooler. The pump is usually driven by the gearbox itself and an electrically driven pump is normally used during start up. Gears are also used on slow and medium speed engines to drive the camshaft and shaft generators.

Gears and gearboxes tend to be taken for granted, yet they are complex systems that require careful attention. The condition of the lubricating oil is particularly important. The oil should also be regularly analysed to check for viscosity, water and wear elements.

On new gearboxes there may be debris left over from assembly. This debris needs to be flushed out. The initial oil charge should be changed after trials and the oil filters should be renewed or cleaned. Smaller gearboxes may have magnetic plugs. The plugs should be inspected and the amount of debris on them noted. Magnetic filters are also fitted and these should be cleaned regularly. If any large particles are found then the gears should be inspected.

When removing covers to inspect gears the covers and their surrounds should be thoroughly cleaned before removal. Ensure nothing has been left on top of the gearbox that could drop inside. This also applies to items that may be in the top pockets of overalls such as pens or notebooks. If the gearbox has to be left with the cover off ensure it is covered with a plastic sheet to prevent dirt ingress. Dirt ingress can be a particular problem in dry-dock where vent fans may be drawing in blasting dust from outside.

When inspecting the gears the engine should be turned slowly and each gear tooth examined. The starting position should be noted to ensure all teeth are inspected. The surface contact marking should be an even width along the tooth length. If wider at one end it may indicate some misalignment. Any damaged teeth should be marked and sketches or photographs of the damage taken for use during future inspections.

It is important that oil is maintained in a clean condition. Oil samples should be sent for analysis at the same time as the main engine. This should include spectrographic analysis to determine wear metals as well as to test for water and viscosity. If wear metals show an increasing trend, ferrographic analysis should be carried out to determine the size and type of wear particles.

The basic failures of gears are due to:

- Overload.
- Bending fatigue.
- Pitting.
- Adhesive wear.
- Abrasive wear.
- Scuffing.
- Corrosion.
- Polishing.
- Cavitation.
- Electrical damage.

Cavitation and electrical damage are rare, however, electrical damage can cause small pits due to ineffective earthing or defective shaft earthing equipment.

Pitting

Pitting is a common cause of gear tooth failure as they are subjected to high contact stresses and many stress cycles. Pitting is a fatigue process which initiates with a crack at, or just below, the tooth surface. The crack will propagate for a short distance and then a piece of metal will break away leaving a pit. The crack generally starts at the surface where the oil film is very thin and there is a likelihood of metal to metal contact. Crack formations below the surface are usually due to inclusions in the metal. Abrasive particles in the oil can also cause pitting as they become embedded in the tooth surface forming stress raisers.

Adhesive or Scuffing Wear

As stated in section 2.2.4., this type of wear is classed as mild if it only causes wear to the oxide layers. Once this level has been breached and metal to metal contact occurs, the wear becomes severe and is termed scuffing. Scuffing occurs if there is a reduced supply of lubricant or the oil film thickness gets reduced due to excessive temperature, incorrect viscosity or overload.

During 'running in' mild adhesive wear occurs as surface asperities are removed. This is quite normal. A gearbox should be operated at reduced load during the running in process to avoid severe wear occurring. Asperities are removed during running in, therefore, the gear teeth become smoother and the wear rate reduces.

Abrasive Wear

Abrasive Wear is caused by contaminants in the oil which may occur due to the following:

- Internally generated due to wear.
- Introduced during maintenance.
- Leaking seals or cover joints.
- Leaking breathers.

Dirt causes scoring of the teeth in the sliding direction and, as previously stated, may become embedded in the teeth and lead to stress concentrations. Breathers on the gear casing should have filters fitted and all seals and joints should be tight to prevent dirt ingress.

Scuffing

Scuffing is severe adhesive wear and occurs when the lubricant film is too thin allowing metal to metal contact to occur. The high load between the gear teeth and the sliding action causes the metal surfaces to weld together and then tear apart. As they separate material is removed from one tooth.

Gear oils should contain Extreme Pressure (EP) additives containing anti-scuffing compounds. Under the action of the high temperature and pressure which occurs between two gear teeth, the additives combine chemically with the metal surface to form a solid layer on the surface. This layer has a lower shear strength than the metal. Compounds containing sulphur, chlorine or phosphorous are generally used. These additives do not work at lower temperatures.

Polishing

In some cases the excessive pressure (EP) additive may cause polishing of the gear teeth causing them to adopt a highly polished appearance. If this occurs a different EP additive could be required and the oil supplier should be contacted.

2.3 Turbochargers and Air Coolers

2.3.1 Turbochargers

The turbocharger consists of a single stage impulse turbine connected to a centrifugal impeller via a shaft. A typical turbocharger is shown in Figure 77.

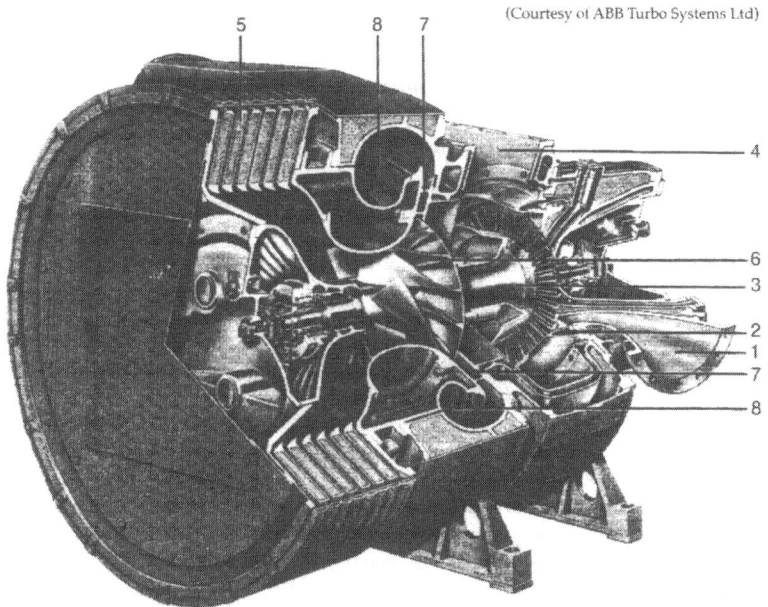

(Courtesy of ABB Turbo Systems Ltd)

Figure 77. Exhaust Gas Turbocharger

The turbine is driven by the engine exhaust gas, which enters via the gas inlet casing (1). The gas expands through a nozzle ring (2) where the pressure energy of the gas becomes converted to kinetic energy which drives the rotor (3). The exhaust gas then passes into the exhaust pipe via the gas outlet casing (4).

On the air side air is drawn in through mesh filters (5), which also act as a silencer, and enters the impeller (6) axially where it is accelerated to high velocity. The air exits the impeller radially and passes through a diffuser (7), where some of the kinetic energy gets converted to pressure energy. The air passes through the scroll shaped casing (8) to the charge air cooler. Labyrinth glands provide sealing on the air and gas sides to prevent gas and oil leakage. Sealing air may also be provided via passageways in the casings.

The bearings can be rolling element or plain journal. Different lubrication systems are used for each. Rolling element bearings will have integral shaft driven oil pumps, which supply oil from a small reservoir. A high volume must be supplied as the oil is also used for cooling the bearing. Rolling element bearings have a series of spring plates around them which are also filled with oil to provide some damping for the bearing. One of the bearings will also act as a thrust bearing.

Journal bearings usually have the oil supplied from the engine crankcase oil system via a header tank. The higher life of journal bearings allows them to be fitted between the impeller and rotor. This makes for easier maintenance as the bearings do not need to be removed when removing the casings.

The gas inlet casing is insulated from the air side and usually water cooled, though some modern designs of turbochargers have no water cooling. Water cooling also cools the exhaust gas at the exit from the turbine, which increases the risk of acid corrosion, especially at lower loads. A non water cooled casing makes for a much simpler casting.

Turbine blades are usually Nimonic 80 material (72 per cent nickel, 18 per cent chrome with other trace elements such as titanium) which has good resistance to creep, fatigue and corrosion. Blade roots are of fir tree shape which, for its size, gives minimum stress concentration at the root. The root is usually a slack fit to allow for differential expansion of the rotor and blade and to assist damping vibration. Damping wire can also be fitted to the blades; this is a loose fit with short lengths of wire for several blades. Modern designs of turbochargers do not tend to have damping wire fitted as blades are a tight fit in the rotor.

Reasons for Turbocharging

- Air charge density is increased allowing more fuel to be burned.
- Improved scavenge efficiency of the engine.
- Reduced thermal load on engine components.
- Increased engine output.

Turbocharging systems fall into two categories, pulse and constant pressure. The system used depends upon the engine type.

Pulse System

This system is used in four stroke engines and utilises the energy of the exhaust gas when the exhaust valve opens to drive the turbocharger. The piping system is more complex than the constant pressure system. Pipes have to run from each cylinder to the turbocharger. The turbocharger gas inlet casing has several inlet ports depending upon the number of engine cylinders. For instance, where nine cylinders are supplying the turbocharger, the gas inlet casing may have three inlets with three cylinders supplying each inlet. This is shown in Figure 78. The exhaust pipe volume has to be kept small to maintain the exhaust gas pressure after it leaves the cylinder.

The high energy of the exhaust gas entering the turbocharger results in it being very responsive to engine load changes. This also means an auxiliary blower is not required.

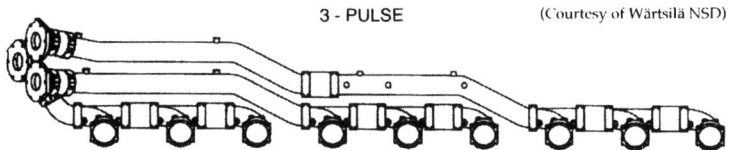

3 - PULSE (Courtesy of Wärtsilä NSD)

Figure 78. Pulse Turbocharged System

Constant Pressure System

This generally applies to two stroke engines. The exhaust gas is discharged from the cylinders into a large manifold where the gas pressure is reduced to below scavenge pressure. The turbocharger operates at maximum efficiency due to the constant pressure of the exhaust gas. The exhaust valve opening point does not depend on a need to supply the turbocharger and allows the exhaust valve to open later, giving a longer power stroke. The constant pressure system is not very efficient at low loads, therefore, an auxiliary blower is required for low load operation.

2.3.2   Turbocharger Surging

The turbocharger has to be correctly matched to the engine to give optimum efficiency. Turbocharger performance is dependent upon the gas angle at entry to the impeller, diffuser and turbine and these are correct at a specific rotor speed. At other speeds the gas angle does not match the blade angle so losses increase. To give better efficiencies at part load the turbocharger can be designed to over-boost the engine. A waste gate can be utilised to achieve the correct boost at full load. This is sometimes used on medium speed engines.

The turbocharger must produce the required scavenge pressure while maintaining an adequate reserve against surging. Surging is the breakdown of gas flow and a reversal of flow from the scavenge space through the diffuser and impeller. Surging is undesirable as it interferes with combustion and more importantly, increases the possibility of thrust bearing failure. Consider the plot of pressure ratio against flow as shown in Figure 79.

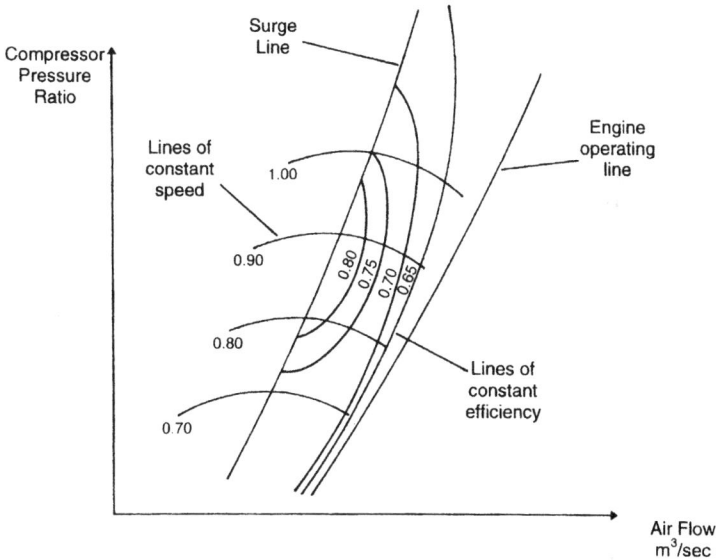

Figure 79. Turbocharger Characteristic Curves

The lines of constant speed show that as the air flow increases the pressure ratio rises and then falls off. Drawing a line through the peaks gives a surge line. To the right of the line the compressor is stable. It can be seen that a reduction in flow gives a rise in pressure. To the left of the line is the unstable region where a reduction in flow gives a reduction in pressure. In the unstable region the turbocharger cannot maintain pressure and the air flow decreases further. Once the delivered pressure falls below the scavenge pressure, flow reversal occurs and the turbocharger surges.

Causes of Surging

- Fouled compressor or turbine, or dirty intake filters. These should be regularly cleaned to prevent a build up of deposits.

- Poor power distribution between the cylinders. Take draw cards to check. May also be caused by rapid change in load or rpm, a particular case is in bad weather when the engine may be racing. A fouled hull will also cause the engine operating line to move towards the surge line, increasing the risk of surging.

- High exhaust back pressure. Usually evident by higher than normal exhaust temperatures. On a pulse turbocharged system if one section of the nozzle ring is choked the cylinders feeding that section will have higher exhaust temperatures.

Causes may be:

- Exhaust valve not opening properly.
- Fouled nozzle ring.
- Fouled exhaust inlet grid.
- Fouled scavenge air cooler.
- Choked scavenge ports.

If surging occurs the engine speed must be reduced. As the most common cause is fouling of the air side, the compressor should be water washed and the air intake filters cleaned. If this does not solve the problem the engine balance should be checked by taking a set of indicator cards.

2.3.3 Turbocharger Deposits and Cleaning

Deposits

The deposits found on turbochargers are:

Compressor: Oil mist, dust, soot, etc. which are all derived from the atmosphere.

Turbine: Fuel derived ashes such as sodium/vanadium compounds, sulphur compounds, metallic ashes such as iron and nickel. If the fuel has a high level of contaminants such as sodium, iron, calcium, magnesium, etc. there may be excessive ash deposits.

Cleaning

Compressor:

A small fixed container is provided which is filled with water to clean the compressor. The water is injected using the air from the turbocharger, so the higher the compressor speed the better the cleaning. Cleaning is carried out at full load and performed once per day. *Solvents should not be used.* The procedure is as follows: (See Figure 80)

- Open the filler on the tank and fill with fresh water only. Secure filler cap and close vent.
- Open the air supply valve A.

-   Open injection valve B and wait for 30 seconds.
-   Close valves A and B and open vent.
-   Check to ensure the tank is empty. If not, there may be a blockage, usually at the bottom of the tank, due to rust and scale, or at the injection nozzle.

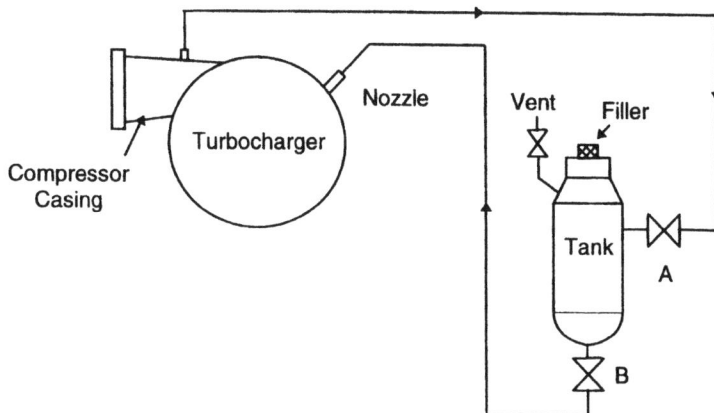

Figure 80. Water Washing Turbocharger Compressor

Turbine:

The turbine can be water washed or dry cleaned.

Water washing

The engine speed must be reduced to reduce the exhaust temperature and prevent thermal shock of the turbine. Once the exhaust temperature is at or below the manufacturer's limit, the turbocharger drain can be opened and fresh water admitted to the turbine casing. Water should be admitted slowly until water appears at the drain, then the water flow can be increased. Washing should continue until the drain water is clear, usually 10 to 15 minutes. The water supply can then be closed off and the drain left open. Once the water has stopped the turbocharger speed can be increased, then the drain closed. This operation is usually carried out on a weekly basis.

Dry cleaning

The turbocharger speed does not have to be reduced when drycleaning. A container is filled with the correct amount of cleaning material, either ground nutshells, activated carbon or small grains of rice. The valve from the container should be opened to blow the material into the turbine casing. After a couple of minutes the valve can be closed. This should be carried out every day or at least every two days.

As with compressor cleaning, no solvents should be used for cleaning the turbine, and if the turbocharger has not been cleaned for a long time it should not be cleaned in service. Deposits may not be evenly removed causing a risk of unbalance. The turbocharger must be opened up and cleaned by hand.

2.3.4 Turbocharger Defects

Table 2 shows a summary of turbocharger problems and possible causes.

POSSIBLE CAUSES \ PROBLEM	Exhaust Gas Temp Before Turbine High	Charge Air Pressure Low	Charge Air Pressure High	Turbocharger RPM Low	Turbocharger RPM High	Lube Oil Pressure Low	Loss of Lube Oil	Slow Runup and Rundown Time	High Noise level	Vibration	Surging
Silencer or Air Filter Fouled	■	■			■						■
Compressor Fouled	■	■			■						■
Turbine Wheel Fouled	■	■			■						■
Nozzle Ring Slightly Fouled	■		■	■						■	
Nozzle Ring Heavily Fouled	■	■		■						■	
Thrust Ring or Labyrinth Damaged							■		■		
Defective Labyrinth Seals							■				
Leaking Seals or Connections							■				
Defective Bearings or Rotor Imbalance	■							■	■	■	
Rotor Rubbing								■	■		
Foreign Bodies in Turbine				■				■	■	■	
Foreign Bodies in Compressor		■						■	■	■	
Damaged Turbine or Compressor Wheel								■	■	■	
Ineffective Sealing Air	■			■							
High Air Inlet Temp	■	■									
Low Air Inlet Temp		■			■						
Fouled Air Cooler	■	■									
Charge Air Inlet Temp High	■	■									
High Lube Oil Temp						■					
High Lube Oil Pressure							■				
Lube Oil Filter Dirty						■					
Deposits on Inlet/Exhaust Valves	■	■									■
High Exhaust Gas Back Pressure	■	■									
Fuel Injection System Requires Attention	■										■

Table 2. Turbocharger Fault Finding

Operation with Turbocharger out of Operation

In all cases the engine manufacturer's instructions should be adhered to.

Operating the engine without a turbocharger will reduce the air supply to the engine and increase the thermal load. The engine rpm will need to be reduced and exhaust temperatures closely monitored to keep them as even as possible.

2.3.5   Air Coolers and Cleaning

Engines are turbocharged to increase the amount of air entering the engine and allow more fuel to be injected. As the turbocharger is compressing the air an increase in charge air temperature will occur. This will reduce the density of the air entering the engine. To reduce this effect the air is cooled which is the function of the charge air cooler. There is also the added benefit that cooling the air reduces the engine thermal load.

Air coolers should be regularly drained as large quantities of water can be removed from the air during the cooling process. In tropical conditions, where the air is very humid, as much as 1000kg of water per hour can be produced.

Maintenance consists of cleaning the air and water sides of the tubes and inspecting the anodes in the water headers. The air side will gradually become fouled with oily deposits, which can be monitored by measuring the air differential pressure. Engine manufacturers usually have limits for the differential pressure indicating when the cooler should be cleaned.

The air side can be cleaned on a regular basis. While the engine is in service small dosing pots can be filled with diluted air cooler cleaner. Only special cleaners that are designed for this should be used. Other solvents should not be used.

Eventually the air cooler will require more thorough cleaning. On larger engines there may be a circulating system for cleaning the air coolers. For smaller coolers these may have to be removed and soaked in a special solvent or cleaned ultrasonically. A typical air cooler cleaning system is shown in Figure 81.

The air cooler cleaner can be circulated in undiluted form and should be circulated for at least 12 hours, depending upon the degree of fouling. The air cooler should be hosed with clean fresh water after cleaning.

On the water side the tubes should be cleaned by brushing through with special tube brushes. This is ideal for removing soft deposits and slime, but if the tube has scale on the inside or hard marine growth, it may require cleaning with a high pressure hose or de-scalant.

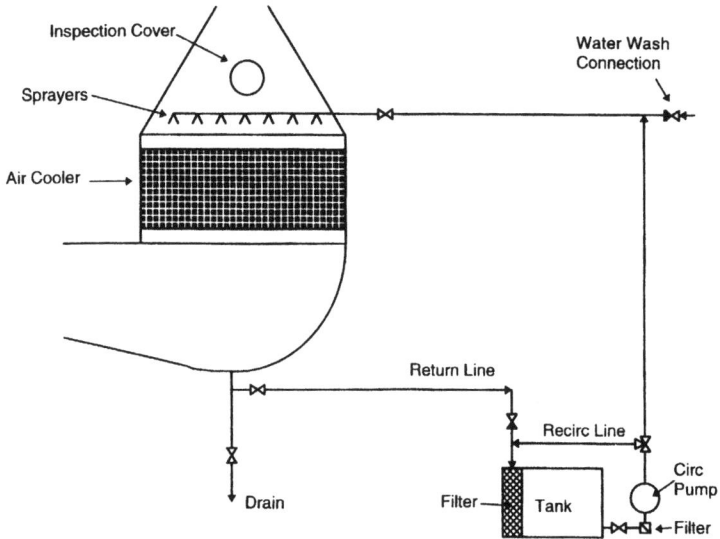

Figure 81. Air Cooler Cleaning System

Leaking tubes can be a problem with air coolers and usually appears in the form of excessive water at drains. It is possible to find the tube leak and plug it while the engine is still running. The procedure for this is as follows:

- Reduce the engine speed. The air after the turbocharger should be less than 70°C.

- Isolate the cooling water and remove the cooler headers.

- Completely cover the tubes at one end of the cooler with a sheet of thick jointing. This can be held in place by the end cover.

- Brush soapy water over the other ends of the tubes. Leaking tubes will produce bubbles, which should be marked.

- Plug leaking tubes with tapered copper or wooden plugs.

Note: If a tube is suspect first make sure the drain water is sea water, as it may be nothing more than condensate from the charge air. Leaking liner seals can also be a source of water, so check the header tank level is not falling.

PROBLEM	CAUSE
Charge air temperature too high	High engine room temperature
	High sea water temperature
	Cooling system out of balance (low flow)
	Fouled air cooler tubes (water side)
	Fouled air cooler tubes (air side)
Charge air temperature too low	Defective cooler controller
	Sea water temperature too low
High pressure drop on air side	Air cooler fins fouled
	Defective pressure gauge
High pressure drop on water side	Fouled air cooler tubes on water side
	High coolant flow
	Defective pressure gauge
External water leakage	Leaking joint on tube plate or pipe connection
Internal water leakage	Leakage between tube and tube plate
	Holed tube

Table 3. Air Cooler Problems

2.3.6   Scavenge Fires

For a scavenge fire to occur the following conditions need to be present at the same time:

- Combustible material must be present in the scavenge spaces. This could be fuel or partially burned fuel, cylinder lubricating oil or carbonaceous material.

- There must be a source of ignition. This is usually blowby between the rings and liner. Blowby can occur due to:

  i. Excessive liner wear.

  ii. Broken piston rings or rings seized in grooves.

  iii. Worn piston rings or ring grooves.

Causes of Scavenge Fires

  i. Excessive engine wear. Worn liners, rings and ring grooves.

  ii. Broken or jammed piston rings.

  iii. Dirty scavenge spaces.

iv. Poor combustion resulting in afterburning. This destroys the cylinder oil film and results in poor ring lubrication and higher instances of wear and/or seizure.

v. Incorrect cylinder lubrication; too little, too much or incorrect timing.

Indications of a Scavenge Fire

i. Scavenge temperature alarm high.

ii. High exhaust temperature.

iii. Loss of engine power and reduction in engine rpm.

iv. Surging of turbocharger(s).

v. Smoke in the engine room and from the scavenge drains.

vi. Black smoke and sparks from the funnel.

vii. Paint blisters on the scavenge doors - in extreme cases.

Action in the Event of a Scavenge Fire

The action taken depends to some extent on the size of the fire, which may be difficult to determine. If paint is blistering on the scavenge trunking, turbochargers are surging and the engine revs are greatly reduced the fire can be considered severe. The engine can even starve itself of air if the fire is large enough.

For small fires:

i. Reduce engine speed to slow or dead slow.

ii. Lift the fuel pump of the affected cylinder(s).

iii. Increase cylinder lubrication. This has to be carefully monitored to ensure it does not feed the fire. If the fire increases then do not increase lubrication.

iv. Keep scavenge drains closed.

v. Monitor the exhaust and scavenge temperatures and let the fire burn itself out.

vi. Once the fire appears to be out, engine revs can be increased cautiously.

vii. Continue to monitor for any signs of re-ignition.

If the fire repeatedly re-ignites after it has been extinguished it may be necessary to open up the scavenge space for inspection after allowing the engine to cool down.

For large fires:

i. Stop the engine and engage the turning gear.

ii. Extinguish the fire with a smothering system, if fitted. If this system has not been fitted, the scavenge trunking should be boundary cooled until the fire has died out. This will prevent heat distortion.

iii. Once the fire is out and the scavenge manifold cooled down, the scavenge space should be opened for inspection and cleaning.

2.4 Fuels and Bunkering

This section is a general overview. For more detailed information the reader is referred to IMarE's Marine Engineering Practice Series, Volume 3, Part 19, *A Practical Guide to Marine Fuel Oil Handling* by Chris Leigh-Jones.

2.4.1 Fuels

Modern marine engines now burn residual fuel and, in many cases, auxiliary engines burn the same fuel as the main engine. Residual fuels, as the name implies, are the residue remaining after all the high quality products have been extracted from the crude oil.

Residual fuels were traditionally 'straight run', that is, produced from the residue from the fractionating column. In the 1970s secondary refining techniques were introduced to extract more distillate from each barrel of crude oil. This resulted in less residue, which was also of poorer quality.

Marine residual fuels are generally blends of residue and distillate mixed to give the required viscosity and density. This blend can vary greatly and thus the quality of fuel is a large variable. The vast majority of residual fuels, however, are treated and burned with no problems.

For a distillate fuel the cetane index is used as a measure of the fuel's ignition quality, however, there is no accurate method to determine the ignition and combustion quality of a residual fuel. The calculated carbon aromaticity index (CCAI) can be used to give some indication of ignition quality, but ignition delay and rate of heat release are also important factors as these affect the rate of pressure rise.

Fuel Properties

Marine residual and distillate fuels are classified according to certain physical and chemical properties. ISO 8217:1996 is the international standard that categories the fifteen residual and four distillate fuel grades.

Viscosity

This is a measure of a fluid's resistance to flow. Viscosity is measured in centistokes (cSt) and often quoted at a reference of 50°C for residual fuels and

40°C for distillate fuels. Many engine manufacturers state that their engines can burn fuel up to 700 cSt @ 50°C. The maximum viscosity which can be handled on board, however, is limited by the heating equipment. Most modern marine engines burn 380 cSt fuel, or IFO 380 as it is termed.

Viscosity is no guarantee of fuel quality and it is a misnomer that lower viscosity residual fuels are of better quality. Knowledge of the fuel's viscosity is required to select the correct transfer and injection temperatures.

Density

This is the relationship between mass and volume (kg/m^3 or kg/l), usually at a reference temperature of 15°C. As with viscosity, density is not a measure of fuel quality but is required to determine the quantity delivered as tanks are calibrated in volume. Density is needed to select the correct gravity disc (if fitted) for the separators.

Flashpoint

This is the temperature at which vapour will ignite when exposed to a flame. The flashpoint of marine fuels should not be less than 60°C, though for emergency equipment located outside the machinery space it can be as low as 43°C.

Pour Point

This is the lowest temperature at which the fuel can be handled before wax crystallisation will prevent flow. Some residual fuels can have high pour points, therefore, knowledge of the pour point is required for setting the correct storage temperature. Fuels should be kept about 5°C above the pour point.

Carbon Residue

This is the tendency of a fuel to form carbon deposits in the absence of air. A fuel with a high carbon residue may lead to increased fouling of pistons and rings. The fuel may also take longer to burn.

Ash

This is the inorganic matter in the fuel and is related to the type of crude oil and the refinery process. Ash consists of the naturally occurring elements in the crude oil, such as nickel and vanadium, and elements that have been introduced during the refinery processes, such as silicon and aluminium from catalytic fines. Ash content may increase due to contaminants in the fuel, for instance salts from sea water contamination. High ash content fuel can lead to increased fouling of combustion components, the exhaust system and turbochargers.

Sediment

This is the amount of insoluble residue in the fuel, such as rust scale, dirt, sand, etc. Sediment is introduced after refining or during storage or transportation. Sediment can also be caused by sludge from an unstable fuel.

Stability

Stability is the resistance of the fuel to chemical breakdown and resulting sludge deposit. An unstable fuel will produce sludge by itself and this may be initiated by heating or moving the oil around the vessel.

Compatibility

If two fuels are blended together to form a homogeneous mixture without forming any sludge, they are said to be compatible. Incompatible fuels can form large volumes of sludge which will block filters, valves and pipes, and quickly choke separators. When these fuels are burnt in the engine, both injection and combustion can be affected. Injectors and rings may become fouled and cylinder components damaged. Compatibility can be tested with a simple spot test of a 50/50 blend of the two fuels.

Calculated Carbon Aromaticity Index (CCAI).

This is presently used as an indication of the ignition quality of a residual fuel and was developed by Shell. It is based on a correlation between viscosity, density and aromaticity. For instance, increasing density with constant viscosity results in a poorer ignition quality fuel. The CCAI can be calculated from a formula or nomograms.

A higher CCAI indicates a lower ignition quality and the limit is about 870. Fuels with a high CCAI may be slow to ignite and then burn rapidly giving a very high pressure rise. This may lead to ring collapse in severe cases. The late ignition also results in afterburning and increased thermal stress on the cylinder components.

Cetane Number and Cetane Index (CI)

This is used to express the ignition quality of diesel oil. The Cetane Number is an indication of the ignition delay, which is the time between injecting the fuel and ignition. A higher cetane number indicates better ignition quality. As the cetane number is derived from engine tests, cetane index is more commonly used and is derived from an empirical formula. As with the cetane number, a low cetane index indicates an increase in ignition delay.

Impurities in Fuel

Water

Water may be fresh or saline and is perhaps the most common contaminant in a fuel. Water can usually be removed by adequate settling and efficient centrifuging. Saline water will lead to an increase in fouling due to the sodium present and will also combine with any vanadium present leading to a risk of high temperature corrosion. Water in the fuel can also cause damage to fuel pumps and injectors.

Sulphur

A high sulphur content is undesirable as it poses a risk of low temperature corrosion. Sulphur also has a low calorific value and therefore lowers the calorific value of the fuel. Sulphur content varies widely but for residual fuels it is usually about 3-3.5 per cent by weight.

During combustion sulphur in the fuel combines with oxygen to form sulphur dioxide (SO_2). Further oxidation results in the formation of sulphur trioxide (SO_3) and vanadium compounds can act as a catalyst for this process. Sulphur trioxide then combines with water vapour to form sulphuric acid (H_2SO_4). This acid vapour will condense on liner walls and, if not for the highly alkaline cylinder oils, would cause corrosion. The temperature at which the acid condenses is called the dew point and it is important that cylinder liner wall temperatures are kept above the dew point temperature.

Vanadium

This occurs naturally in the crude oil and the amount varies depending on the source. Venezuelan crude oils have very high levels. Vanadium is oil soluble and cannot be centrifuged out.

In the combustion process vanadium oxidises to form various compounds that combine with sodium to form further compounds which are highly corrosive in the liquid state. Compounds such as vanadium pentoxide and sodium vanadates attack the protective oxide layer on high temperature components, such as exhaust valve seats, exposing the base metal which is then attacked. Furthermore the various compounds can combine to form eutectics which have stiction temperatures as low as 300°C. This is the temperature at which deposits will adhere to metal surfaces (see Figure 82).

Sodium

This is usually associated with sea water contamination. One per cent sea water in the fuel equates to about 100ppm sodium. If sodium is present without water it is oil soluble and cannot be removed. Sodium salts from water, such as sodium chloride, will rapidly foul exhaust valves and turbochargers. Sodium may also combine with sulphur in the fuel to form sodium sulphate (Na_2SO_4), which is a corrosive slag at high temperatures.

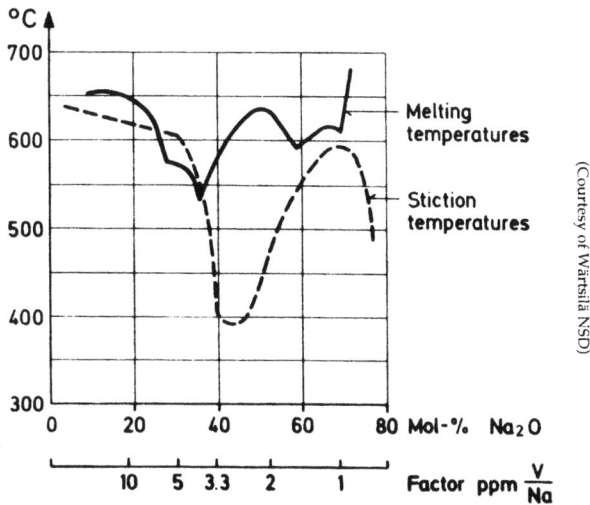

Figure 82. Sodium/Vanadium Equilibrium Diagram

Catalytic Fines

Catalytic fines are the small particles of catalyst that are used in a catalytic cracking plant. Catalytic cracking is a secondary refining process used to increase the yield of lighter products from the crude oil. The catalysts used are aluminium silicates and are generally recovered after use, although they may be carried over into the residue in some instances.

The 'fines' are very hard abrasive particles which cause damage to fuel pumps and injectors. Efficient separation should reduce any catalytic fines to an acceptable level.

2.4.2   Bunkering

Bunkering is a routine operation for ships but, despite this, it has potential dangers. The risk of a spill can be virtually eliminated provided the whole bunkering operation is planned beforehand with a tank plan and checklist.

The Bunkering Plan

Ordering the correct quantity of fuel involves several parties, the owner/ manager, charterer and the Chief Engineer. The vessel will usually be informed well in advance of the quantity of fuel to be bunkered and the ship's staff can then set out the bunkering plan.

The bunkering plan basically consists of the following:

- A list of the tanks to be filled.
- The order in which they will be filled.
- Start and finish tank ullages or soundings.
- Tank quantities.
- Responsibilities of ship's staff involved in the bunkering operation.

The number of tanks filled will depend on the amount of fuel to be bunkered. The officer responsible for the trim of the vessel should be informed of how many tanks are being filled so the vessel's draft and trim can be determined before departure from port.

To avoid incompatibility problems the fuel should always be bunkered into empty tanks, where possible.

Tank gaugings can be either ullages or soundings. Once the number of tanks to be filled has been decided the tank tables must be used to determine the start and finish ullages or soundings. Ullages and soundings should not be used together as this can lead to confusion. Ullages tend to be more accurate as there may be debris, sounding weights, etc. at the bottom of the sounding pipe which could give a false sounding.

Tank tables are in volume whereas fuel is supplied in tonnes, therefore, an approximate figure must be used for density to determine the tank tonnage. Some tables have tonnage conversions at several densities, however, over a third of residual fuels have densities between 985.0 and 991.0 kg/m^3 (or 0.9850 and 0.9910 $kg/litre$), so using a figure between these should be close. A figure of 850.0 kg/m^3 is suitable for distillate fuels and diesel oil.

Maximum bunkers is generally taken as 98 per cent. If not taking maximum, although some tanks are required to be full, try and keep to 90 per cent capacity. This will reduce the risk of a spill. First the volume at the required per cent tank capacity, e.g. 90 per cent, should be determined then tank tables can be used to note the ullage or sounding. Apply the approximate density to give a tonnage. This can be carried out for each tank until the total tonnage is reached. The information should then be listed in a table, as shown in Figure 83, as part of the vessel's bunker plan. The tanks should be listed in the order of filling.

Tank tables also include corrections for trim and in some cases heel. The officer responsible for the trim of the vessel should again be consulted to determine the approximate trim when the vessel is bunkered. If the trim is expected to vary widely and maximum bunkers are being taken, the tank tables will need to be consulted and the bunker plan amended accordingly.

TANK	START ULLAGE	FINISH ULLAGE	QUANTITY TONNES
1C DB	15.50M	13.60M	200
2P DB	14.00M	12.05M	180
2S DB	14.00M	12.05M	180

Figure 83. Bunker Tank Filling and Quantity List

Once the bunkering plan has been completed a list should be made of personnel involved in the bunkering operation, along with their responsibilities. This is a United States (US) Coastguard requirement for vessels bunkering at US ports. It is good practice and should be used wherever a vessel is bunkering.

The Bunker Checklist

There are many tasks to carry out prior to and during bunkering. These may be forgotten with all the other activities going on while in port. Using a checklist ensures no task is missed. This is also a requirement for the US Coastguard. Table 4 shows a typical basic checklist that can be modified to suit any particular vessel.

Before bunkering, the supplier may ask the vessel to sign bottle labels and other paperwork. No receipts should be signed at this stage. Request a copy of paperwork from the barge, showing grade and density, and check that they are correct. If viscosity and density are not acceptable then refuse to accept the fuel. Only accept figures on headed paper. Note that some densities may be quoted at 20°C or specific gravity (s.g. @ 60/60°F) or API gravity @ 60°F. The bunker barge soundings/ullages should be witnessed by ship's staff before and after bunkering in case of a dispute over quantity.

During Bunkering

Under normal circumstances only one tank should be filled at a time. Trying to gauge more than one tank at a time increases the risk of an overflow. As one tank gets near to its final ullage, the second tank should be partially opened up to allow the first tank to continue to fill at a slower rate.

If the tanks are filling too quickly the barge should slow down the pumping rate. Also, if taking maximum bunkers the tanks should be filled to 90 per cent and then topped up at a slower rate at the end. For maximum bunkers it is best to use a figure of 95 per cent, rather than 98 per cent, to minimise the risk of a spill. For a 500 tonne tank the difference between 90 and

95 per cent is 25 tonnes. This will only take 6 minutes if the pumping rate is 250 tonnes per hour. *It is not worth having a spill trying to squeeze in a few extra tonnes.*

PRIOR TO BUNKERING	
BUNKER PLAN COMPLETE	
CHECK ON BOARD COMMUNICATION SATISFACTORY	
OVERFLOW TANK EMPTY	
DECK SCUPPER AND SAVEALLS PLUGGED	
ADEQUATE LIGHTING AT BUNKER STATION AND SOUNDING POSITION	
NO SMOKING NOTICES POSITIONED	
OPPOSITE SIDE MANIFOLD VALVES CLOSED AND BLANKED	
BARGE MOORINGS SECURE	
HOSE CONNECTED AND SEALED SECURELY	
HOSE ADEQUATELY SUPPORTED	
ESTABLISH METHOD OF COMMUNICATION WITH BARGE	
RED FLAG OR LIGHT AT MASTHEAD	
CHECK BARGE PAPERWORK FOR GRADE AND DENSITY	
CONFIRM WHICH GRADE TO BE PUMPED FIRST AND PUMPING RATE	
ALL APPROPRIATE FILLING VALVES OPENED	
OPEN MANIFOLD VALVE	
AT START	
INSTRUCT BARGE TO START PUMPING AT SLOWER RATE	
CHECK MANIFOLD CONNECTION FOR LEAKS	
CHECK HOSE IS SUPPORTED WITH WEIGHT OF FUEL IN IT	
OPEN SAMPLING VALVE AT MANIFOLD	
WHEN SATISFIED THAT TANK IS FILLING INCREASE PUMPING RATE	
AT FINISH	
ONCE BARGE ADVISES FINISH, CLOSE MANIFOLD VALVES	
CLOSE ALL TANK FILLING VALVES	
TAKE FINAL ULLAGES/SOUNDINGS	
RECORD TIME OF START & FINISH BUNKERING, TANKS AND QUANTITY IN EACH, IN ENGINE LOG BOOK AND OIL RECORD BOOK	

Table 4. Sample of Typical Bunker Checklist

Soundings or ullages should be taken regularly during bunkering and even more so when the tank is nearing completion. Many vessels have tank gauges but, unless these can be relied upon as accurate, they should only be used as a guide.

During the bunkering procedure a continuous sample should be taken in a clean, dedicated container. If using a manual sampler or sampling cock it will need to be checked regularly to ensure the oil has not stopped flowing.

The fuel temperature should be checked. The supplier will also advise the delivery temperature. A temperature higher than this could lead to a shortfall after applying the correction to the density figure.

After Bunkering

A full set of tank soundings and ullages should be taken, even if the tanks have gauges. The soundings should be corrected for trim and heel to determine the total volume delivered. This should then be converted to tonnes by applying the density. The density on the bunker receipt should be corrected for the temperature of the fuel. The receipt density will usually be at 15°C so residual fuels will be heated prior to pumping on board. Remember to also correct the densities of distillate fuel. Although these are not heated, the ambient temperature may be above or below 15°C.

As a general rule, for each degree rise in temperature the density should be reduced by 0.64 kg/m³. Volume correction factors may also be used instead, so the density is taken to be at 15°C and then multiplied by a correction factor.

If there is a shortfall in delivered quantity the vessel should issue a note of protest to the supplier and sign the bunker receipt for the quantity 'as determined by vessel's tank soundings'.

Ensure the 'vessel retained sample' and the 'suppliers sample' are retained for any future reference. Samples taken during bunkering should be sent for laboratory analysis.

2.5   *Preparing for Sea & Arrival in Port*

2.5.1   Preparing for Sea

The following assumes the engine is being prepared from cold.

24 hours prior to departure

- Ensure fuel tanks are at correct levels and if necessary transfer/purify fuel. Open up heating to the service tanks to warm the fuel to at least 40°C.

- Ensure the jacket water header tank level is satisfactory and ensure all valves in the system are open.

- Start the jacket water circulating pump and check the pressure. Occasionally jacket water control valves can stick so these should be changed to manual control and operated over their full range to check for free operation. Then change back to automatic control. Check the returns to the header tank are all flowing freely.

- Preheat the jacket cooling water gradually to at least 60°C. If the engine is cold this should be carried out over a 24 hour period.

6 hours prior to departure.

- Start the lubricating oil pumps for the main engine, camshaft system, and crosshead, as applicable. Check the sight glasses for oil flow. Check all pressures are satisfactory. If the oil is very cold it may need to be warmed through before starting the pumps.
- Check the oil levels in the main engine sump tank, turbochargers (if fitted with a separate oil system, start the oil pump and check oil levels), and shaft bearings.
- Start the lubricating oil purifiers about 6 hours before departure to warm through the oil. If the oil is very cold this should be done before starting the main lubricating oil pumps.

1 hour before departure (This is usually when 'Standby Engines' is rung)

- Test the Bridge and local control steering gear by operating from port to starboard on each pump in turn, then with two pumps running (in conjunction with deck officer). Check all oil levels.
- Start the controllable pitch (CP) propeller pumps (if applicable) and test the pitch control. Check pressures and oil levels.
- Start the main gearbox lubricating oil pump (if applicable) and check pressures and oil levels.
- Start the standby generator and connect to the switchboard.
- Operate the cylinder lubricators and turn the main engine over with the turning gear. Ensure the indicator cocks are open and that no water or fuel issues during turning.
- Drain the main air receivers of any water.
- Switch on the oil mist detector and observe the operation through one cycle.
- Start the sterntube lubricating oil circulating pump (if applicable) and check the header tank levels.
- Take readings of the fuel oil settling and service tanks and cylinder lubricating oil daily service tank.

15 minutes before departure

- Open the main air receiver valves and the main start air valve.
- Ensure all the main air compressors are ready for use.
- Blow over the main engine on air, ahead and astern, with the indicator cocks open.

- Where turbochargers have integral pumps, check through the sightglasses that they are supplying oil. After long periods idle or after overhaul, these may need priming, often achieved by removing the oil filling plug above the pump and pouring oil over the pump.

- Close the indicator cocks, scavenge drains, air cooler drains, turbocharger drains, etc.

- Set the auxiliary blower to automatic.

- Start the main engine fuel oil booster pump (if on distillate fuel) and check fuel oil pressure. Prime the fuel injectors if the engine has been stopped for a long period with the fuel pumps switched off.

- Start the fuel valve cooling water pump (if applicable) and check pressure and temperature.

- Inform the Bridge that the main engine is ready for use and put the engine on Bridge control.

- Check the standby units are set ready to operate.

Checks once the engine is running

- Check the turbochargers for oil supply.

- Check the shaft bearing oil circulation.

- Check all the exhaust valves are operating.

- Check all the cylinder lubricators are working.

- Check all exhaust temperatures are satisfactory to ensure all cylinders are firing. Exhaust temperatures may vary widely at low loads and firing can be checked by putting a hand on each injector fuel pipe cladding to feel the 'pulse'. If this is weak and the exhaust temperature is low, the injector may need priming. Another cause may be a sticking fuel pump suction valve, particularly if the engine was left on residual fuel in port.

- Feel all the air start supply pipes just before the air start valve. If one pipe is getting hotter than the others, the valve is probably leaking. This may be due to a problem with the valve itself or, if the valve was changed in port, the seal between the head and valve may be damaged.

- Check all pressures and temperatures.

- When 'Full Away On Passage' is rung, take the following readings:
  - Fuel oil service and settling tank levels.
  - Cylinder lubricating oil daily service tank level.
  - Main engine RPM counter.
  - Main engine fuel oil flowmeter.

- Engine speed should be increased slowly from manouervring speed to full sea speed. This is usually carried out by an automatic engine load program but if this is not fitted or is defective, the engine speed should be increased to full sea speed gradually.

2.5.2   Preparing for Arrival in Port

Standby is usually rung just before the engine needs to be manoeuvred. Preparations need to be made before this time.

1 hour before arrival

- Commence reducing from sea speed to manouervring speed.
- Commence changing the main engine over to diesel oil (if required). The decision to do this depends upon the type of engine, length of stay in port and the type and duration of any maintenance work to be carried out.
- Start a second generator engine and parallel with the running generator.
- Ensure the power supply to the bow thruster and deck machinery is available.
- Start the standby steering gear unit (usually done from the Bridge).
- Raise pressure in the auxiliary boiler with the steam stop valve closed. This will then be ready for operation when the exhaust gas boiler is shut down.
- Blow down the main air receivers and ensure the standby air compressor is ready for use.
- Open air to the main engine and drain the line of any water. Note: this should always be open during UMS operation.
- Ensure the auxiliary blower is set to automatic. This should always be left on automatic during UMS operation.
- Test the engine operation astern. After several days or weeks at sea it is not unknown for control valves to stick. Coming up to the berth is not a good time to find out the engine will not go astern. For vessels with CP propellers, these should also be tested astern. When the test has been carried out it should be noted in the engine logbook. This is mandatory on all vessels prior to entering US waters.
- When 'End Of Passage' is rung, record the fuel oil settling and service tank and cylinder lubricating oil daily service tank readings, the engine RPM counter and the main engine fuel flowmeter reading.

Arrival in Port

Once in port and 'finished with engines' has been rung, the engine can be shut down. Record the fuel oil settling and service tank and cylinder lubricating oil daily service tank readings.

The duration in port will determine how many systems are shut down. If the stay is to be for more than a few days then the cooling systems may be shut down and the engine should be changed over to diesel oil. Modern engines have recirculation systems which allow the engine to remain on heavy oil which, depending on the type of work that needs to be carried out on the engine, can be left running.

- Change over to Control Room Control.

- Shut the main air start valve and drain the line.

- Turn off the auxiliary blower.

- Open the indicator cocks.

- Engage the turning gear.

- Open the scavenge, turbocharger and air cooler drains.

- Stop the camshaft and crosshead lubricating oil pumps (if separate).

- If the main engine is on diesel oil stop the fuel oil booster and fuel valve cooling pumps.

- If deck power is not required the standby generator can be taken off the board and shut down.

- Stop the main sea water pumps. Note: this depends upon the system on board the vessel and may require auxiliary sea water pumps to be started.

- After 15-30 minutes shut down the main lubricating oil pumps.

- If the port duration is only a few hours, open up the jacket water preheating.

2.6   Watchkeeping and UMS Operation

Most vessels now operate with an Unmanned Machinery Space (UMS), therefore, watches only tend to be kept when entering and leaving port. UMS operation requires that the machinery alarm and monitoring systems are fully operational. If alarms are not working then watches must be kept at all times.

The routines involved for the watchkeeping engineer and UMS duty engineer are similar, as UMS is essentially a 24 hour watch. Traditionally, the watchkeeper would enter the engine room prior to taking over the watch, and carry out an inspection of all machinery. The watchkeeper who was handing over would discuss any problems that had occurred during the watch drawing attention to any items that may need particular attention.

The same still applies with UMS operation and the oncoming duty engineer should have a thorough inspection of the machinery spaces before taking over. Any problems should be brought to the attention of the present duty engineer. Remember that when you are on duty or watch you are responsible. If something goes wrong while you are in charge it is no good trying to blame someone else.

The principles to be observed during watchkeeping are covered in the International Convention on Standards, Training and Certification of Watchkeepers (STCW) code, Annex 1, Pt A-VIII/2 and this is well worth reading.

If problems occur while you are on duty that you are unsure about, no matter how insignificant they may seem, you should call the Second or Chief Engineer.

Before the machinery space can be left in unmanned mode at night the duty engineer should carry out an inspection. This will ensure nothing has been missed and that all important items have been checked. A UMS checklist should be used and all items checked off and signed by the duty engineer. The list should follow an orderly route around the machinery spaces. A typical list is shown in Table 5.

The check list is only a guide to show some of the items that should be included, however, one of its purposes is to ensure all areas of the machinery spaces are inspected.

Once the checklist has been completed the duty engineer should sign the sheet. This sheet should then be kept as a permanent record. The Bridge watchkeeping officer should be informed when the engine room is being left unmanned and the dead man alarm should be activated. This is a safety feature and, once the duty engineer enters the engine room, the dead man alarm system needs to be reset every 20 or 30 minutes. If the engineer has an accident and cannot reset the alarm, this will sound on the Bridge to alert the watchkeeper.

During unmanned periods of operation the main starting air supply to the main engine must be left open. This ensures that the Bridge have full control of the engine. Safety equipment such as fire pumps, that can be started from the Bridge, should also be left ready for operation.

If alarms occur during UMS periods they should be recorded in an alarm log book. Details of the date and time of alarm, what the alarm was and any action taken, should all be recorded.

UMS operation relies on a correctly functioning alarm system. It is therefore important to have an ongoing routine of alarm and automatic shutdown testing. These tests should be recorded and, along with the alarm log book and UMS checklists, provide a record for the Classification Society Surveyor when carrying out the UMS survey.

CHECK POINT	CHECKED	REMARKS
Funnel Smoke	✓	
Funnel Door Shut	✓	
Economiser Visual Inspection	✓	
Blow Down Gauge Glasses	✓	
Blow Down Level Switches	✓	
Incinerator Off	✓	
Drain Incinerator Sludge Tank	✓	
Main Engine Header Tank Levels	✓	
Distilled Water Tank Levels	✓	
Workshop Machinery Off	✓	
Gas Bottles Shut Off	✓	
Main Engine T/C Oil Levels	✓	
Visual Inspection of M/E Cyl Heads	✓	
Inspect Air Cooler and Scav Trunk	✓	
Inpect Engine Middle Platform	✓	
Jacket Water Cooler		
Lube Oil Cooler		
Economiser Circ Pump		
Fresh Water Generator Salinity		
Chemical Dosing Tank Level		
Distillate Pump	✓	
Main Air	✓	
	✓	
	✓	
	✓	
...is on	✓	
...tor Sump Level	✓	
Generator T/C Oil Level	✓	
Generator Governor Oil Level	✓	
Generator Pedestal Bearing Oil Level	✓	
Emergency Air Receiver Pressure	✓	
Steering Gear Hyd Tank Level	✓	
Steering Gear Pump	✓	
Auto Greaser Tank Level	✓	
Steering Flat Bilge Well	✓	
Control Room Alarm Panel Clear	✓	
Remote Boiler Level Gauges	✓	
All Standby Machines Selected	✓	
440V and 220V Earth Test	✓	
Duty Engineer Selector Switch	✓	
Read and Signed C/E Standing Orders	✓	
Dute Engineer		Date

Example

Table 5. UMS Checklist

Provided thorough checks are made of the machinery spaces, and alarms regularly checked for operation, unmanned periods of operation should be trouble free.

Chief Engineer's Standing Orders

The Chief Engineer will issue standing orders in the form of continuous instructions to be followed by the engineering officers to allow safe and efficient operation of the main and auxiliary machinery. Some shipping companies have their own standing order books, but one can easily be made up.

Some of the information that may be included in standing orders is as follows:

- General procedures for UMS and watchkeeping operations.
- Use of oil record book.
- Waste disposal.

- Protective clothing to be worn.
- When to call Chief Engineer, e.g. standby, any problems, engine slowdown, etc.
- Remarks to enter in log book, e.g. times of standby, end of passage, stops at sea, etc.
- Any particular operational requirements, e.g. space heaters required in cold climates.
- Weekly running of emergency equipment.
- There should also be space for additional orders to be entered on a daily basis, if required, and the book should be signed by the engineer officer taking over the watch or duty.

2.7 Performance and Condition Monitoring

Engineers are on board ship to ensure the safe and efficient operation of machinery. Safe and efficient operation of any individual item of machinery, whether it is a small domestic hot water pump or the main engine, requires regular maintenance.

Maintenance requires manpower and time and in some instances these are not always available. Gone are the days when ships carried a large complement of engineers. A 1960s edition of *Marine Steam Engines & Turbines* by Fox and McBirnie gives a list of watch-keeping duties which includes the 10th engineer. Clearly things have dramatically changed over the last few decades.

Many ships now operate with only three engineers on board and short periods in port provide little time to carry out maintenance. Performance and condition monitoring techniques are therefore being used more and more to provide information as to when machinery needs overhauling, and to keep the engine at its optimum for maximum fuel economy.

There are various maintenance schemes that can be followed and these can be depicted using the block diagrams in Figure 84.

1. Preventative or Scheduled Planned Maintenance

Maintenance is carried out at predetermined intervals, either calendar or hours based. Maintenance is undertaken irrespective of machinery condition.

2. Corrective/Breakdown Maintenance

Repair of a particular item when it fails. This is not an ideal situation and is only suitable for inconsequential items which can only be maintained on a repair by replacement basis.

3. *Condition Based Maintenance*

This is dictated by the performance or physical state of the machine and is undertaken when operating conditions have deteriorated below a satisfactory standard.

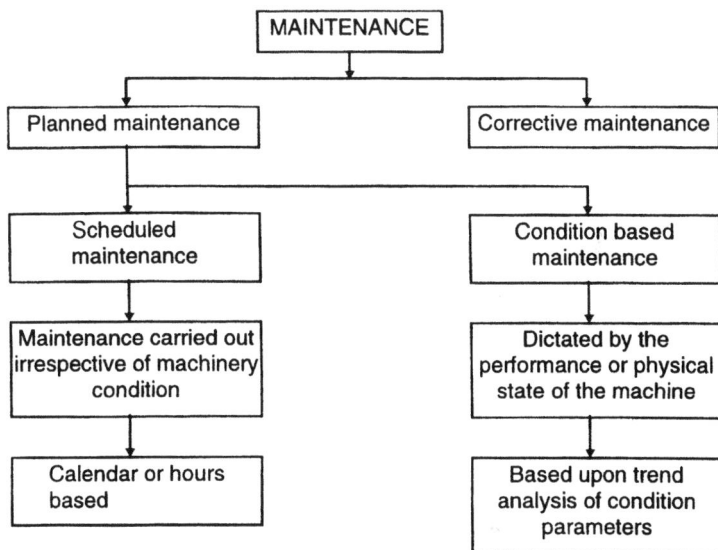

```
                    ┌──────────────────┐
                    │   MAINTENANCE    │
                    └──────────────────┘
         ┌───────────────────┴───────────────────┐
┌──────────────────────┐              ┌──────────────────────┐
│ Planned maintenance  │              │ Corrective maintenance│
└──────────────────────┘              └──────────────────────┘
┌──────────────────────┐              ┌──────────────────────┐
│     Scheduled        │              │  Condition based     │
│    maintenance       │              │   maintenance        │
└──────────────────────┘              └──────────────────────┘
┌──────────────────────┐              ┌──────────────────────┐
│ Maintenance carried  │              │   Dictated by the    │
│ out irrespective of  │              │ performance or       │
│ machinery condition  │              │ physical state of    │
│                      │              │ the machine          │
└──────────────────────┘              └──────────────────────┘
┌──────────────────────┐              ┌──────────────────────┐
│ Calendar or hours    │              │ Based upon trend     │
│ based                │              │ analysis of condition│
│                      │              │ parameters           │
└──────────────────────┘              └──────────────────────┘
```

Figure 84. Maintenance Schemes

2.7.1   Performance Monitoring

This entails regular monitoring of the combustion conditions, engine pressures and temperatures to determine engine condition. Combustion monitoring is carried out to ensure the engine is operating at peak efficiency to ensure maximum fuel economy. Fuel quality can have a great effect on engine operation and combustion monitoring allows timing adjustments to be made to take poor quality fuel into account. Engine fuel pump timing is usually carried out statically, with the engine stopped, whereas performance monitoring systems allow the timing to be checked dynamically.

Combustion monitoring is usually carried out by taking a set of draw or power cards (see Indicator Diagrams 2.2.7). Not all engines have the facility to take power cards, however, electronic combustion monitoring equipment is fitted on many new vessels.

Early electronic indicators were basic oscilloscopes which took signals from a sensor on the engine flywheel and a pressure transducer fitted to the indicator cock. Modern systems use computers and performance monitoring software to display more parameters and have built in diagnostic capabilities. Sensors can also monitor fuel pump pressure, performance of turbochargers and air coolers, scavenge and exhaust temperatures, and cylinder pressures and temperatures. Parameters may also be compared with model curves under ideal conditions and trends can then be shown graphically and printed off. A typical system block diagram is shown in Figure 85.

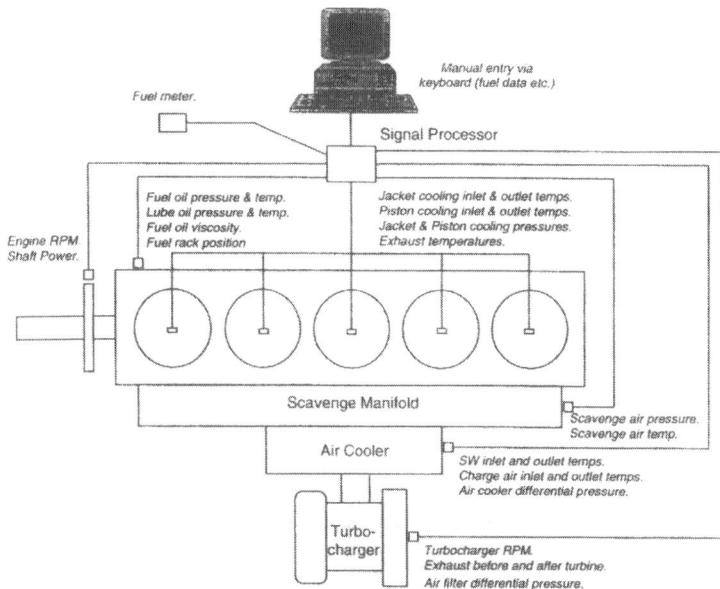

Figure 85. Performance Monitoring System Block Diagram

Condition Monitoring

Condition monitoring can be defined as 'an assessment on a continuous or periodic basis of the mechanical condition of machinery, equipment and systems from the observations and/or measurements of selected parameters'.

Some of these measurement methods are:

- Human senses, such as sight, smell, touch, etc.
- Pressure/temperature monitoring.
- Vibration analysis.
- Oil analysis.

- Piston ring/liner wear (SIPWA).
- Ultrasonics.
- Strain gauges.
- Fibre optics.

Traditional watchkeeping operations used human senses for condition monitoring as the watchkeeper would inspect all machinery visually and feel for any hot bearings. The watchkeeper could also 'listen' to bearings by placing a screwdriver on the bearing cap and against the ear. The Chief Engineer would also monitor trends visually in the engine log book. The main problem with human senses is the component is usually in an advanced state of deterioration when it is detected - it is too late when the paint on the bearing housing is observed to be blistering with heat.

Pressure and temperature monitoring have already been mentioned under performance monitoring and perhaps the other frequently used methods on board ship are oil and vibration analysis. Static inspections of components such as turbochargers using boroscopes are also used and strain gauges may be found on large tankers and bulk carriers for hull stress monitoring.

Oil Analysis

All machinery with moving parts will wear. This wear debris will be transported away by the lubricating oil. Under normal operation the amount of debris will be small and should not give cause for concern.

Oil analysis involves taking a sample of oil from the machine and sending it to a laboratory for analysis. The key to getting meaningful results is to ensure the sample taken is representative of the oil in the system. The sample should be taken from a position where the oil is in full flow and not from filter or cooler drains. The lubricant should also be at its normal operating temperature and a few litres should be drained from the sampling line before filling the sample bottle. Samples should not be taken after fresh oil has been added to the system.

Samples should be taken at three monthly intervals and analysed for viscosity, water, insolubles, base number and wear metals, as shown in Tables 6 and 7. It is important to monitor trends in wear metals and if these show an increasing trend then a more in-depth analysis, such as ferrography, should be carried out. This determines the size and type of wear debris as the wear particles have distinctive characteristics such as rubbing wear, fatigue chunks, severe wear and cutting wear.

Typically samples are taken from the main and generator engines and sterntube system, however, other systems, such as steering gear and deck hydraulic systems are just as important. The majority of breakdowns in hydraulic systems are due to a deterioration in the cleanliness of the oil, either due to water or wear and contaminant debris. Particle counts can be used to determine the oil's cleanliness level.

Contaminants in a hydraulic system can lead to three types of failure.

1. *Catastrophic failure* - This is where a component ceases to operate. If the damage is severe the oil will be contaminated with debris and more components in the system could become damaged.

2. *Intermittent failure* - This is perhaps the most common problem and is caused when dirt or wear debris is large enough to prevent a valve from seating or causing an obstruction to a small orifice. The dirt may eventually wash away and the system will function normally. This type of intermittent fault results in an unreliable system, with ship's staff wondering when the next breakdown will occur.

3. *Degradation failure* - Wear and erosion of components increases clearances and damages valve seats and seals. This causes imprecise control and a reduction in efficiency. If wear is occurring this will escalate and eventually cause catastrophic failure.

TEST	CONDITION	POSSIBLE CAUSE
Viscosity	Increase	Insolubles Oxidation Fuel dilution with heavy fuel grade Incorrect lubricant grade Water contamination
	Decrease	Fuel dilution with light fuel grade Incorrect lubricant grade
Water	Increase	Leaking cooler Condensation Header tank exposed to elements Leaking gland (Sterntube)
Insolubles	Increase	Combustion products Wear debris Extraneous dirt Defective filters/separators
Base Number	Increase	Contamination with cylinder oil Incorrect lubricant grade
	Decrease	Normal due to additive depletion Rapid depletion indicates a problem - may be insufficient alkalinity in new oil, water in intake air, poor combustion
Elements	Increase	See table 7
	Decrease	See table 7. Usually applies to additive elements which deplete over a period of time

Table 6. Lubricating Oil Tests

ELEMENT	POSSIBLE SOURCE
Aluminium (Al)	Bearing alloys (Al/Sn)
	Piston skirts/Crowns
Iron (Fe)	Piston rings
	Cylinder liners
	Crank and camshafts
	Gears
	Rust (Water present)
Tin (Sn)	Tin based white metal bearings
	Bearing overlays (Pb/Sn)
Copper (Cu)	Bearing alloys (bronze/brass bushes)
	Piston rubbing bands
	Piston rod gland rings
	Bearing backings
	Copper pipes/coolers (in hydraulic systems)
	Mechanical seal faces
Chromium (Cr)	Chrome plated piston rings and liners
Lead (Pb)	Lead based white metal bearings
	Bearing overlays (Pb/Sn)
Nickel (Ni)	Residual fuel contamination
	Nimonic components
Calcium (Ca)	Lubricant additive
Magnesium (Mg)	Lubricant additive
	Sea water contamination
Phosphorus (P)	Lubricant additive
Zinc (Zn)	Lubricant additive
	Paint coatings
	Galvanised components
Sodium (Na)	Sea water contamination
	Residual fuel contamination
	Lubricant additive
Silicon (Si)	Sand/silica contamination via air intake or sea water
	Cast iron components
	Lubricant additive
Vanadium (V)	Residual fuel contamination

Table 7. Wear and Contaminant Elements

Vibration Monitoring

All machines vibrate. Vibration should be at a low level if machines are correctly balanced, aligned, and fastened down; and if bearings, gears and drive belts are in good condition. Should any of the above begin to deteriorate the vibration level will increase. Defects also have characteristic vibration signatures that can be used to identify the cause of the vibration.

Vibration monitoring can therefore be used to determine the condition of a machine. Overhaul should be carried out when the vibration reaches a pre-determined level. Some Classification Societies accept vibration monitoring readings when surveying rotating machinery and, providing the readings show a satisfactory trend, the machine can be surveyed without being opened up.

Vibration is usually measured in velocity (mm/s) and acceleration (m/s^2 or g's). Readings are taken on the machine, usually closest to the bearings, in the vertical, transverse and axial directions. Measurements should be taken monthly and trends monitored. If the vibration level at one point starts to increase, measurements can be taken more frequently and a harmonic analysis carried out.

The overall vibration level of a machine is made up of vibration at different frequencies and amplitudes. For instance, gears will produce high frequency, low amplitude vibration, whereas, unbalance is at lower frequencies (machine RPM) and of high amplitude. There may also be vibration from misalignment, loose holding down bolts and other sources such as the main engine. The vibration meter 'sees' all these and combines them into an overall value. Most modern vibration monitoring systems now have the ability to analyse the different frequencies. A typical vibration signature is shown in Figure 86.

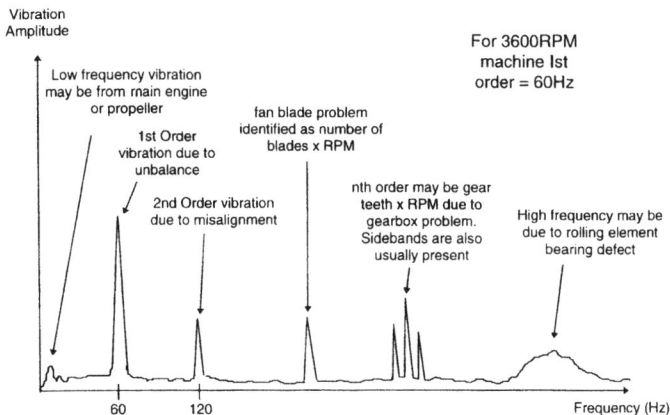

Figure 86. Vibration Frequency Spectrum

The vibration signature or spectrum can be used to help determine where the cause of the problem may be. The highest peak is usually at machine RPM or 1st order, usually due to unbalance which is the most common cause of machinery vibration. The frequency in hertz is equal to machine RPM/60, so 1st order for a machine running at 1800 RPM is 30Hz. Table 8 shows some of the common causes of vibration and their characteristics.

Vibration Frequency	Order	Amplitude	Cause
1 x Machine RPM	1st	Large in radial direction	Unbalance
1 x Machine RPM	1st	Large in axial direction	Axial misalignment
2 x Machine RPM	2nd	Radial. 2nd order highest	Parallel misalignment
0.5 x Machine RPM	$^1/_2$		Oil whirl on high speed machines
No. of gear teeth x RPM	nth		Gear teeth x RPM indicates gear meshing problem. Belt speed x RPM indicates belt problem
n x RPM	nth		Rolling element bearings may not be multiple of RPM and spike energy should be used.

Table 8. Causes of Vibration

3 Auxiliary Machinery

This chapter covers pumps, coolers, centrifugal separators, fresh water generators and air conditioning and refrigerating plant. Sewage plant and oily water separators are covered in Chapter 6.

3.1 Pumps

There are many types of pump installed on board ship and these may be direct drive, belt drive or driven through a gearbox.

3.1.1 Positive Displacement Pumps

Positive displacement pumps are capable of pumping air and can generate high pressures. These pumps are therefore fitted with a relief valve on the discharge side before the discharge valve. It is important that the discharge valve is fully open and discharge pressure is often regulated with the relief valve. Positive displacement pumps can be categorised as follows:

- Reciprocating.
- Rotary - screw, gear, vane, rotor (mono).
- Rotary / Reciprocating i.e. swash plate pump.

Reciprocating Pumps

These are usually motor driven and have a reduction gear to the crankshaft drive wheel. These pumps are most often used as bilge pumps and perhaps only one or two will be found on board. Despite their high efficiency - they have the highest efficiency of all positive displacement pumps due to their low leakage - maintenance is high due to the large number of moving parts.

The discharge pressure is also pulsating, so an air reservoir must be fitted on the discharge side to dampen out the pulsations. The discharge rate is quite small but high discharge pressures can be generated and the flow is regulated by throttling the suction valve. As with all positive displacement pumps, it is self priming.

Maintenance

Maintenance entails lubricating the bearings which can be rolling element bearings with a grease nipple or, on larger pumps, plain bearings which may have oil cups that need to be filled regularly when the pump is in use. Bucket rings are normally ebonite or PTFE and these need to be replaced after a while. Valves also need overhauling, although this mainly consists of cleaning.

PROBLEM	CAUSE	ACTION
No discharge	Discharge valve closed	Open valve
	Suction valve closed	Open valve
	Suction strainer clogged	Clean strainer
	Damaged or sticking suction or delivery valves	Remove valves and inspect
	Relief valve stuck open or spring broken	Overhaul valve
Reduced discharge	Damaged or sticking suction or delivery valves	Remove valves and inspect
	Suction strainer clogged	Clean strainer
	Relief valve stuck open or lifting	Overhaul valve or adjust setting
	Worn bucket rings	Overhaul pump
	Worn pump cylinders	Overhaul pump
Excessive Motor Load	Discharge valve closed	Close valve
	Piston rod glands too tight	Slacken gland packing
	Worn bearings	Overhaul pump
	Worn gear drive	Overhaul pump

Table 9. Reciprocating Pump Fault Finding

Gear Pumps

These usually consist of two gears, one drive gear and the other driven, although some pumps may have one drive and two driven gears. As there is direct meshing of the gears the fluid being pumped must be a lubricant. This type of pump is simple and maintenance is minimal, even more so if bearing bushes are fitted as opposed to rolling element bearings.

Gear pumps are relatively trouble free and most problems tend to be with the relief valve which may get dirt trapped underneath the seat if the valve lifts. Another problem is that pressure often gets regulated by the relief valve and the spring can eventually weaken and possibly break.

Screw Pumps

Unlike a gear pump the screws do not mesh. Instead they are driven by synchronising gears. It is important to ensure the screws are correctly assembled after overhaul. If the synchronising gears are not marked, they should be marked before removing the screws.

Problem	Cause	Action
No discharge	Discharge valve closed	Open valve
	Suction valve closed	Open valve
	Suction strainer clogged	Clean strainer
Reduced discharge	Suction strainer clogged	Clean Strainer
	Pump drawing air	Check tank contents, shaft gland leaking, leak in suction line
	Relief valve setting Incorrect	Adjust valve
	Worn gears/screws	Overhaul pump
Motor overload	High discharge pressure	Check valves are open Cold oil requires heating
	Gland packing tight	Slacken packing
	Pump out of alignment	Realign
	Worn bearings in pump or motor	Replace
Excessive pump vibration	Misalignment	Realign pump and motor
	Cavitation	Check pump is not drawing air and suction pressure is not excessive
	Oil is cold	Heat oil to min pumping temp
	Safety valve 'chattering'	Adjust or overhaul valve
	Worn bearing	Replace
	Foreign matter in pump	Open up and remove

Table 10. Gear and Screw Pump Fault Finding

3.1.2 Centrifugal Pumps

These pumps can only handle fluids of a relatively low viscosity. The discharge pressure is much lower than a positive displacement pump, however, the discharge rate is much higher. For higher discharge pressures, such as boiler feed pumps, multi-stage pumps are required. Centrifugal pumps cannot pump air, therefore, a priming pump is required unless there is a positive suction head.

Impellers can be single or double entry. With double entry the axial thrust is balanced out. Water enters the impeller axially at low velocity and leaves the impeller radially at high velocity. The volute casing has an increasing cross sectional area which converts the kinetic energy of the water into pressure energy. Some characteristics of these pumps are shown in Figure 87.

The discharge rate is controlled by throttling the discharge valve. It can be seen from the graph above that the motor takes minimum current when the discharge valve is closed.

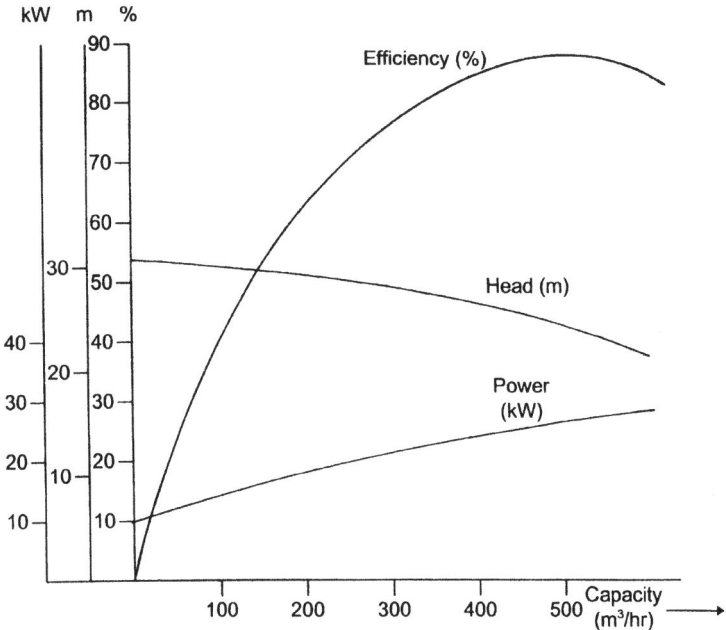

Figure 87. Centrifugal Pump Characteristics

Maintenance

There are very few moving parts so maintenance is mainly concerned with inspecting the impeller and checking seal ring clearances. Impellers are subject to corrosion and erosion due to cavitation, particularly pumps that are used for sea water. The clearances should always be checked against the manufacturers recommendations (see Table 11).

3.1.3   General Pump Maintenance

Alignment

When overhauling pumps it is important to correctly align the motor and pump. Alignment of the coupling can be carried out with a dial gauge for proper accuracy, as shown in Figure 88.

   In most cases the use of a ruler or straight edge and feeler gauges is adequate. The coupling bolts should first be removed and the straight edge placed on the coupling, as shown, in four positions A, B, C and D. The feeler gauges should be used to measure the gap between the couplings at the same positions. The alignment should be adjusted so these readings are the same.

Problem	Cause	Action
No discharge	Discharge valve closed	Open valve
	Suction valve closed	Open valve
	Suction strainer clogged	Clean strainer
	Faulty priming pump	Check operation
	Blocked connection between priming pump and casing	Unblock pipe
Reduced discharge	Suction strainer clogged	Clean strainer
	Pump drawing air	Check tank contents, shaft gland leaking, leak in suction line
	Faulty priming pump	Check operation
	Clogged impeller	Remove and clean
	Impeller vanes wasted away. Excessive seal ring clearance. (Loss of performance over long period)	Overhaul pump
Motor overload	Gland packing tight	Slacken packing
	Pump out of alignment	Realign
	Worn bearings in pump or motor	Replace
Excessive pump vibration	Misalignment	Realign pump and motor
	Cavitation	Check pump is not drawing air and suction pressure is not excessive
	Worn bearing	Replace

Table 11. Centrifugal Pump Fault Finding

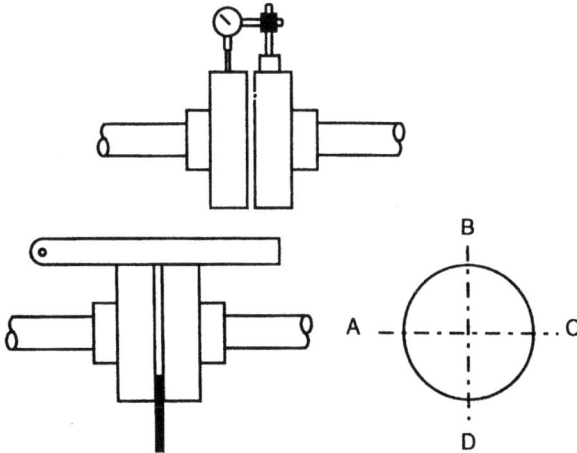

Figure 88. Coupling Alignment

To align belt driven pumps a straight edge is again used, placed across the two pulleys as shown in Figure 89. The correct sized belts must be fitted and spare belts kept stored flat rather than hung up on hooks.

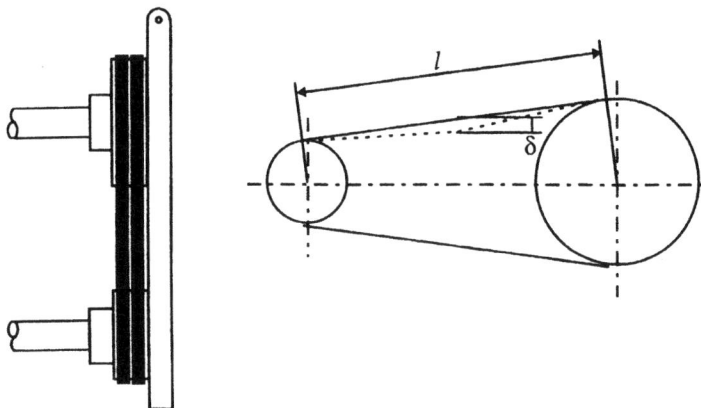

Figure 89. Drive Belt Alignment

The belts must also be of sufficient tension otherwise they may slip and reduce the pump speed and belt life. $\delta = 0.016 \times 1$

Pump Glands

Many pumps now have mechanical seals which are very expensive and should be treated with great care. After removal they should be kept in a clean place. The two seal faces should be carefully cleaned by washing them in the same clean fluid that is being pumped. Avoid wiping the seal faces as any scratches could cause the seal to leak. Conventional gland packing should be replaced at the time of pump overhaul.

- Ensure the gland stuffing box is clean and all old packing has been removed.
- Wrap a turn of packing around the shaft and make it slightly larger than the shaft circumference.
- Cut the lengths of packing with a sharp knife to give a clean butt joint.
- Insert each turn of packing into the stuffing box, offsetting the butt joints by 120º.
- Put in sufficient turns so that the gland ring nuts will fit fully on the studs.
- Fit the gland ring and tighten the nuts hand tight.

When the pump is first run the gland will be leaking. The nuts should be tightened equally half a turn at a time to reduce the leakage. This should be done gradually. Do not try to stop the leakage by tightening the gland down in one session. This would cause the gland to overheat. The packing would then quickly lose its lubricating properties and the gland would leak permanently and the only cure would be to repack it. A very small leakage is desirable to ensure the packing is lubricated. After tightening down it may be possible to fit another turn of packing.

3.2 Centrifugal Separators

Principle of Separation

If we first consider gravity separation, as occurs in a settling tank, over a gradual period solids such as sludge, dirt, etc., will settle out at the bottom of the tank. The heavy liquids, such as water, will settle out above the solids and the lighter liquids, such as oil, will be at the top of the tank. The application of heat will speed up the separation process.

Separation occurs due to the different specific gravities, or densities, of the liquids and solids.

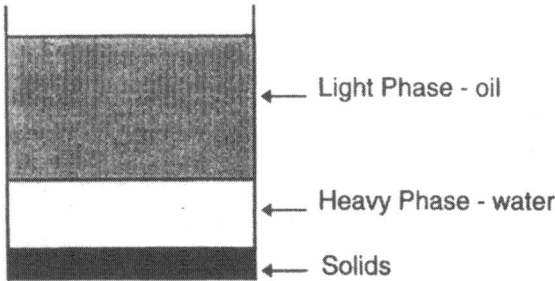

Light Phase - oil

Heavy Phase - water

Solids

Figure 90. Gravity Separation

In the above case it is gravity that is responsible for the settling out and this takes quite a while. If the gravitational force is increased, by using centrifugal force, the separation effect is much greater. This is shown in Figure 91 by imagining the settling tank on its side and rotated.

The centrifugal separator consists of a bowl containing a set of discs, stacked one on top of the other. The dirty oil flows down the distributor and up through the holes in the discs. This is shown diagrammatically in Figure 92.

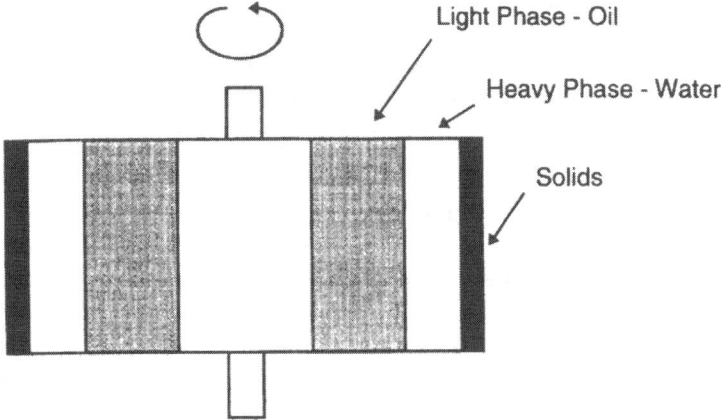

Figure 91. Centrifugal Separation

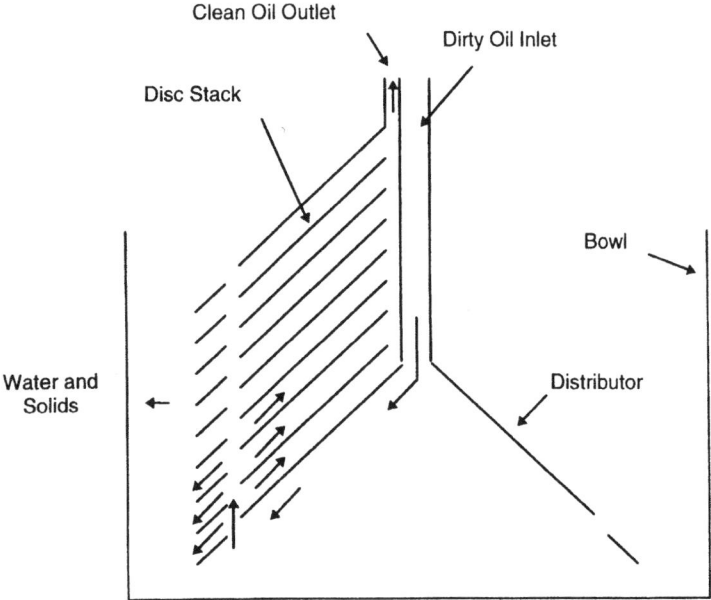

Figure 92. Separator Disc Stack

Considering a solid particle between two discs

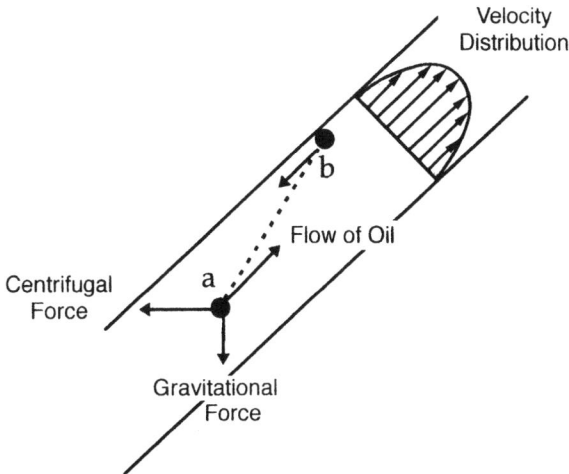

Figure 93. Particle Flow within Disc Stack

The particle is subjected to centrifugal force and a force due to the velocity of the fluid (see Figure 93). The gravitational force is relatively small and has little effect. The resulting flow of the particle will be from 'a' to 'b', until it hits the underside of the disc. At this point the velocity of the fluid is at a minimum and centrifugal force dominates. The particle will then move down the underside of the disc and be deposited at the bowl periphery.

Position of the Oil/Water Interface

In a conventional separator operating in purifier mode it is important that the oil/water interface is in the correct position to allow efficient separation (see Figure 94). This is controlled by means of a gravity disc or regulating ring. There are also other factors that affect the position (see Figure 96). Separator manufacturers provide nomograms to determine the correct gravity disc, based upon oil density and separating temperature. It should be remembered that the difference in densities of oil and water are greater at higher temperatures, therefore, the separating temperature should be as high as possible, preferably 90-95°C.

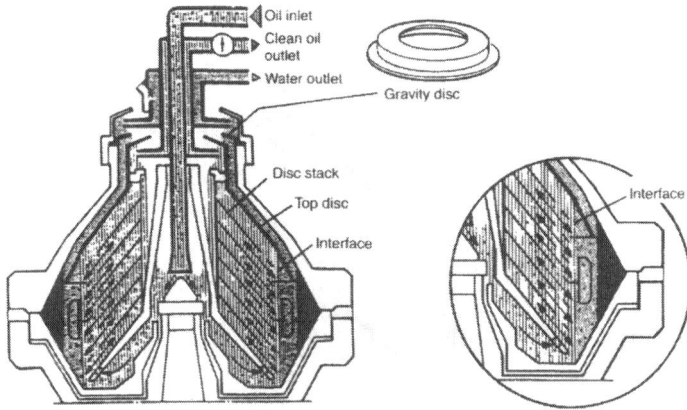

Figure 94. Alfa Laval Purifier Showing Correct Interface Position

The purifier comes supplied with a number of gravity discs with different inside diameters. The optimum position of the interface is between the edge of the disc stack and the outside edge of the top disc. This can be checked visually by looking at the underside of the top disc, shown in the following diagrams (Figure 95).

In a purifier, if the oil is admitted first it will simply discharge from the water side. To prevent this a water seal needs to be applied first. The sealing water should be applied slowly until it is seen discharging from the water side. At this point the sealing water should be closed off and oil admitted. If sealing water has been applied too quickly there is a risk of water passing into the oil side. Occasionally it may be necessary to determine the correct gravity disc by trial and error. This should be done by fitting the largest gravity disc first then gradually reducing the size of disk until the purifier no longer loses its seal, i.e. oil no longer discharges from water side.

Clean oil is clear of water.
Water discharge is clear of oil.

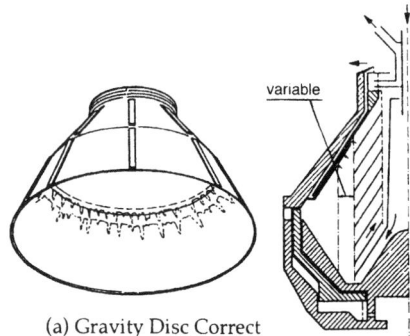

(a) Gravity Disc Correct

Figure 95. Effect of Different Interface Positions

*Clean oil is free of water.
Discharging water contains oil
or purifier loses seal.*

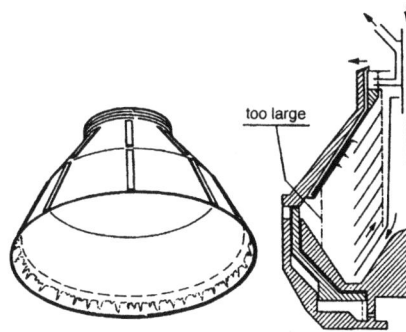

(b) Gravity Disc Too Large

*Clean oil contains water.
Water discharge is clear of oil.*

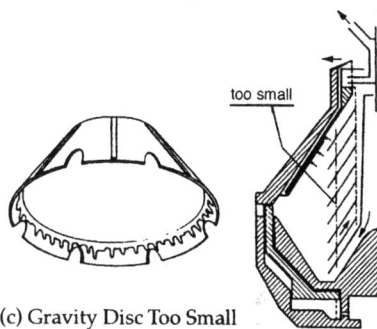

(c) Gravity Disc Too Small

Figure 95. Effect of Different Interface Positions

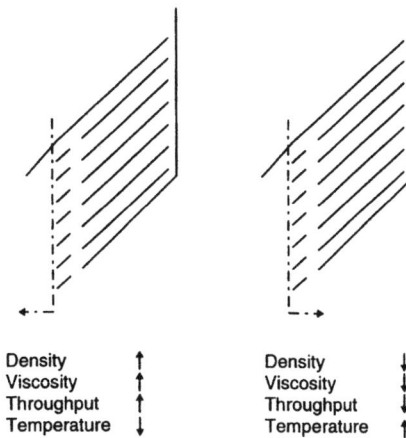

Density	↑	Density	↓
Viscosity	↑	Viscosity	↓
Throughput	↑	Throughput	↓
Temperature	↓	Temperature	↑

Figure 96. Factors Affecting Interface Position

Factors Affecting Separator Efficiency

- *Density difference between water and oil.* Heating gives a greater differential.

- *Position of oil/water interface.* Once water enters the disc stack the separator efficiency is reduced as oil cannot flow along the full surface of the disc.

- *Sludge discharge frequency.*

- *Viscosity of oil.* The lower the viscosity the lower the drag force on dirt particles. Viscosity is reduced by heating.

- *Throughput.* For fuel oil separators the throughput should be as slow as possible to maintain engine demand. Lubricating oil separators are sized to the engine manufacturer's recommendations, and the throughput will depend on the amount of dirt entering the system. In general, for slow speed engines, aim to treat the complete oil charge three to four times in 24 hours, and eight times for medium speed engines.

Purifier or Clarifier

Separators can be operated as either a purifier or clarifier. A clarifier only separates sludge and consequently no water seal is required. The gravity disc gets replaced by a clarifier disc with a smaller inside diameter.

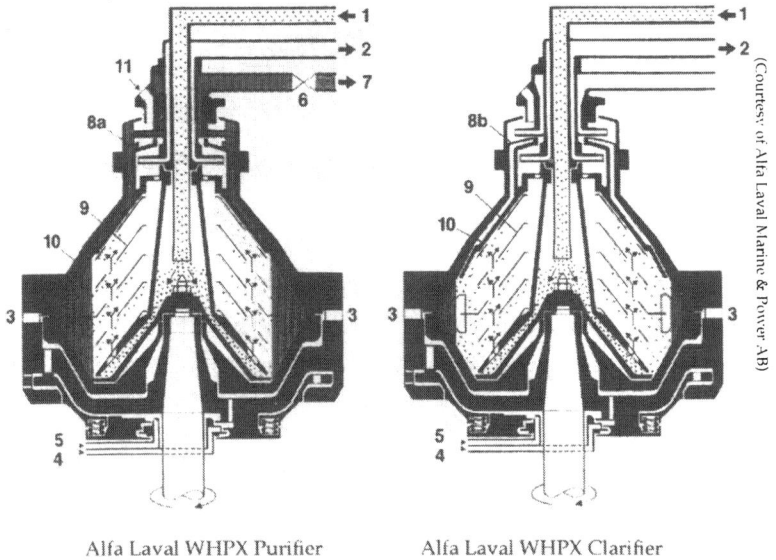

Alfa Laval WHPX Purifier   Alfa Laval WHPX Clarifier

(Courtesy of Alfa Laval Marine & Power AB)

Figure 97. Purifier and Clarifier Bowls

Key

1. Oil inlet	7. Water outlet
2. Oil outlet	8a. Gravity disc
3. Sludge outlet	8b. Clarifier disc
4. Make-up and closing water inlet	9. Disc stack
5. Opening water inlet	10. Top disc
6. Water outlet valve	11. Inlet for water seal & displacement water

Limitations of conventional water seal purifiers

-   Treatment of fuels above 991 kg/m^3 is not possible.

-   Optimum separation depends upon selecting the correct gravity disc, separation temperature and throughput.

-   The oil loss incurred during sludging of the clarifier is relatively high.

The above led to the development of separators which operate without the need to change the gravity disc for different density fuels, such as Alfa Laval's 'Alcap' system and Westfalia's 'Secutrol' system. Some of the benefits of these systems are as follows:

-   Optimum separation efficiency is maintained at all times as water does not enter the disc stack.

-   There is greater flexibility as fuels up to 1010kg/m^3 can be treated.

-   Less supervision is required as there is no gravity disc to change.

-   There are fewer components, therefore, less maintenance.

The Alcap Separator

Dirty oil is continuously fed to the separator bowl. The separator essentially operates as a clarifier, but water as well as solids are separated. Sludge and water are discharged at set time intervals and any excessive water will be automatically drained between sludge cycles.

When the fuel contains little water the sludge discharge is controlled by the timer, where time periods for maximum and minimum sludge cycles are set. During the sludge cycle conditioning water is added to soften the sludge. Displacement water is then added to displace the oil and minimise oil loss during sludging. As the displacement water approaches the disc stack, the water transducer in the discharge line will detect traces of water and initiate the sludge cycle.

If the water content in the bowl builds up gradually the water will be detected by the transducer as the oil/water interface approaches the disc stack. If this occurs after the minimum sludge cycle time, a sludge cycle will be initiated.

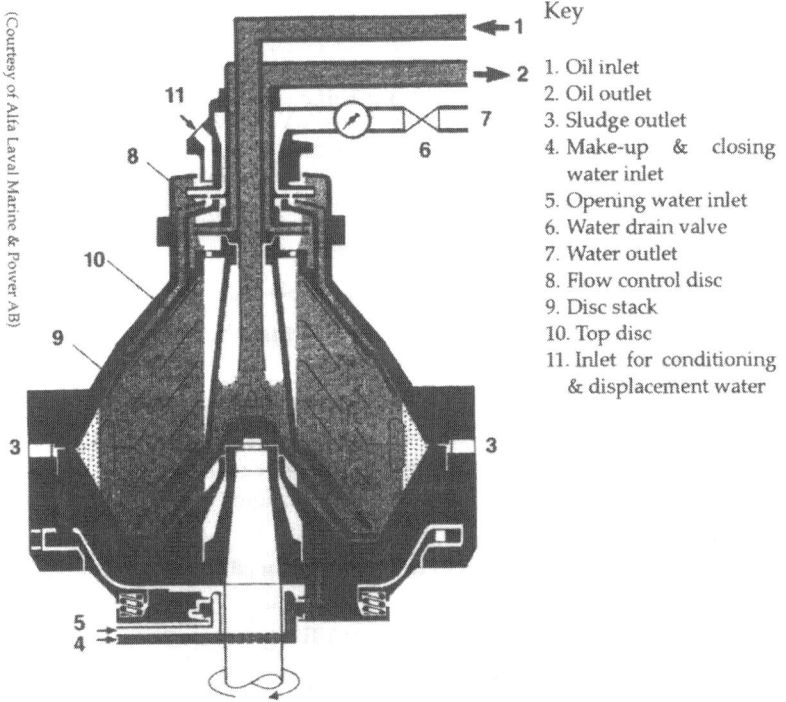

(Courtesy of Alfa Laval Marine & Power AB)

Key

1. Oil inlet
2. Oil outlet
3. Sludge outlet
4. Make-up & closing water inlet
5. Opening water inlet
6. Water drain valve
7. Water outlet
8. Flow control disc
9. Disc stack
10. Top disc
11. Inlet for conditioning & displacement water

Figure 98. Alfa Laval FOPX Separator

If the water in the fuel is high and the transducer detects water before the minimum sludge cycle time, the water drain valve will open to drain the excessive water from the bowl. In extreme cases, where this does not reduce the bowl water content a sludge cycle will be initiated.

Self Discharging Separator Bowl (See Figure 99)

Closed Bowl

The sliding bowl (6) is forced upwards by the pressure of the closing water in space (8). The operating slide (2) is held up by the springs (7). Plugs (10) in the operating slide seal the drain holes (9) in the bowl.

Bowl Opens for Discharge

Opening water (15) is supplied into the space above the operating slide (2). The force exerted by the water overcomes the spring force and the operating slide is pressed downwards. The drain holes (9) are then opened allowing water to drain from space (8) out through the nozzle (3). The sliding bowl (6) then moves downward and the bowl opens, discharging through the ports (1).

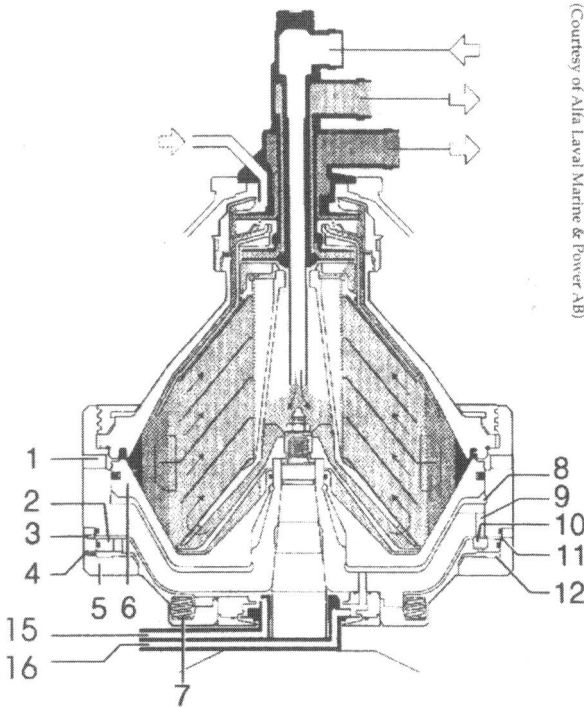

Figure 99. Sliding Bowl Operation

Bowl Closes After Discharge

Water leaving space (8) flows into the opening chamber (11). Channels in the operating slide (2) allow this water to flow into the closing chamber (12) as well. This allows the hydraulic forces to balance out and the force from the springs (7) to predominate, closing the operating slide (2). Now the drain channels are closed, the closing water (16) forces the sliding bowl (6) to move upwards, closing the bowl.

Separator Maintenance

Although separators may be self cleaning, it does not mean that they never need to be opened up and cleaned. The quality of the fuel and the throughput determine the intervals at which the separator bowl needs to be opened up for cleaning.

When cleaning, all the discs should be removed and cleaned. If the sliding bowl assembly has also been removed the sludge chamber should be cleaned.

The sliding bowl does not need to be opened up every time the disc stack needs cleaning but it will need to be cleaned occasionally. All passages and bores for the operating water should be cleaned and the spring lengths checked. Nylon seal rings should be inspected and, provided they are not damaged, do not need to be replaced. It is a good idea to replace other seal rings while the bowl is stripped down.

When reassembling, however, all rubber seal rings should be lubricated with silicone grease. Ensure the bowl spindle and lock ring have been greased with molybdenum disulphide grease. Prior to fitting the bowl assembly check the operating water passages are clear by turning the water on. If the bowl spindle has been removed for any reason the bowl height must be checked when refitting the spindle.

Gearbox oil should be changed at regular intervals and the gears inspected for any signs of undue wear.

(Courtesy of Alfa Laval Marine & Power AB)

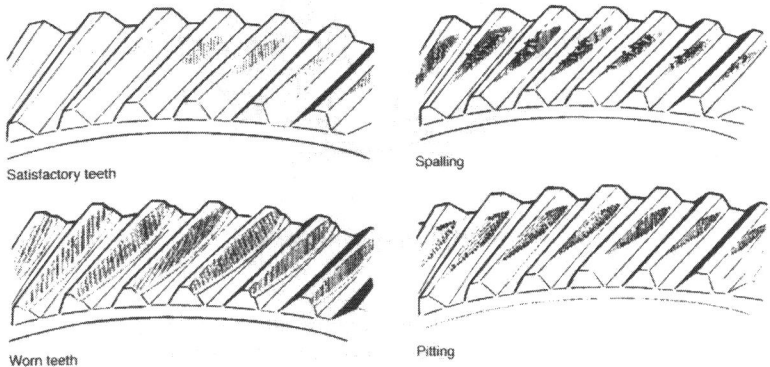

Satisfactory teeth

Spalling

Worn teeth

Pitting

Figure 100. Separator Worm Wheel Gear Teeth Condition

The brake and clutch pad linings should be checked for thickness. Other routine operations include cleaning the oil suction strainer and the strainers on the operating water system.

PROBLEM	CAUSE	ACTION
Bowl speed low or slow to build up speed	Brake still on	Let brake off
	Clutch shoes worn	Renew shoes
	Clutch friction surfaces oily	Remove and clean
	Insufficient clutch shoes	Add another shoe and check speed
	Bowl full of liquid or sludge	Clean bowl
	Bowl height incorrect	Adjust
Separator vibration	Bowl out of balance due to build up of solids	Clean bowl
	incorrect assembly	Check assembly
	Low disc stack pressure	Check number of discs
	Spindle bearing springs weak or broken	Inspect
	Worn bearings	Inspect
	Worn gears	Inspect
Bowl does not close or fails to close properly	Ports in base of bowl are clogged	Remove bowl and clean ports
	Sliding piston water outlet clogged	Remove sliding piston and clean outlet
	Sliding piston not free to move fully	Dismantle and clean
	Worn seal ring on sliding piston	Replace seal ring
	Seal ring on top of bowl is damaged	Replace seal ring
	Operating water does not close off fully	Check solenoid valves
	Bowl is too high	Check bowl height
Oil discharges on water side	Insufficient sealing water	Check pressure and flow
	Incorrect gravity disc	Check nomogram
	Reduced oil temperature	Raise temperature
	Bowl not properly closed	See under problems
	Oil supplied too quickly	Adjust feed valve operation
Oil discharges on sludge side	Bowl not properly closed	See under problems
Water discharges on oil side	Incorrect gravity disc	Check nomogram
	Too much sealing water supplied	Adjust timer for solenoid valve
	Oil supplied too quickly	Adjust feed valve operation
	Bowl failing to open	See under problems
	Water outlet clogged with sludge	Open up and clean
Reduced or no oil feed	Feed pump relief valve lifting	Adjust valve
	Suction strainer blocked	Clean strainer
	Feed pump shear pin failed	Overhaul pump

Table 12. Separator Fault Finding

3.3   Heat Exchangers

Heat exchangers are either shell and tube type or plate type. The majority of coolers fitted on board ship today are of the plate type. Engine air coolers and heaters, however, are usually tube type.

Plate type

- Total strip down allows easy access and more thorough cleaning.
- Damaged plates easily renewed.
- Leaking joints do not cause cross contamination.

Tube type

- Less expensive.
- Easy to locate leaks.
- Leaks cause contamination.
- Spare tube stack often necessary.
- May be difficult to clean.

Cleaning shell and tube type heat exchangers and the associated problems are covered in Chapter 2 under Section 2.3.5.

Tube Type Heat Exchangers.

Materials

Shell　　　Cast iron or mild steel.

Waterbox　Usually cast iron with a rubberised protective coating for sea water service.

Tube plate　60/40 brass.

Tubes　　Aluminium brass, cupro nickel or copper depending upon the service. Zinc anodes are also fitted in the water boxes of sea water coolers. These should have a good connection to the waterbox otherwise they will be ineffective. Ferrous sulphate injection systems are sometimes used to protect against corrosion. This deposits a thin film of iron on non ferrous surfaces which minimises galvanic action. Tube failure is generally due to one of the following:

- General wastage.
- Impingement attack (entrained air and turbulent flow).
- Marine deposits (form organic acids on decomposition which cause pin holes).
- Erosion due to suspended solids, especially in estuarial waters.

The tubes must be free to expand. This can be achieved in several ways.

- To have one fixed tube plate and one floating tube plate with seal rings, e.g. a lube oil cooler.

- Both tube plates fixed with the tubes free to expand through sealed ferrules in one tube plate, e.g.a steam condenser.
- A single tube plate with U tubes, e.g. a compressor intercooler or feed heater.

Plate Type Heat Exchangers

These are built up from a series of rectangular corrugated plates separated by gaskets. The two fluids flow on opposite sides of each plate. A fixed header at one end carries all the pipe connections. At the free end is a removable header. Figure 101 shows a typical plate type heat exchanger.

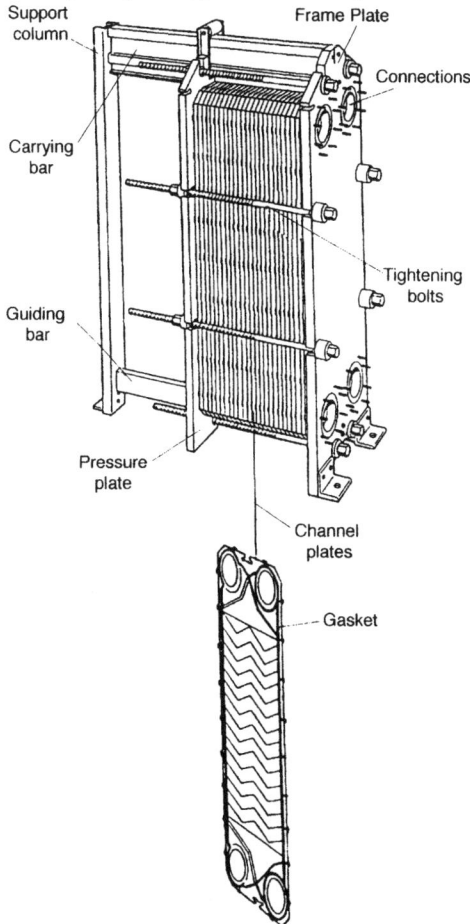

Figure 101. Plate Heat Exchanger Construction

The plates can be made from cupro-nickel, stainless steel or titanium alloy. The material used depends on the service requirements.

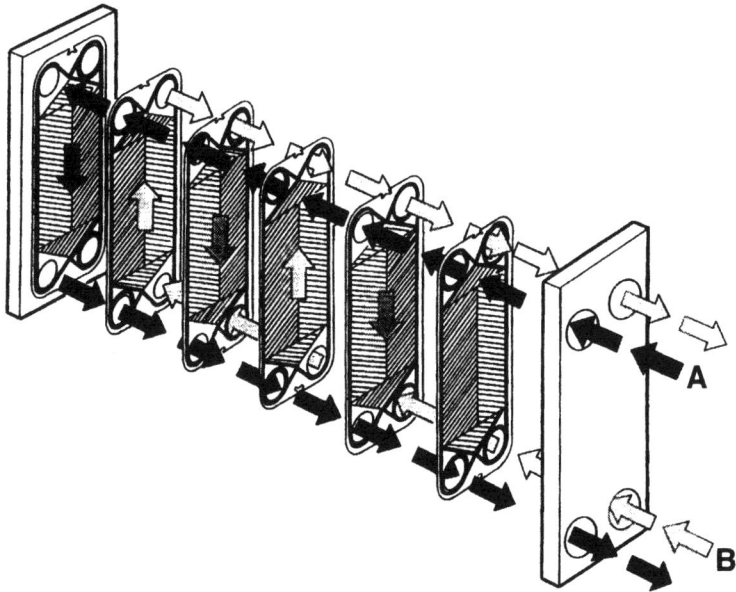

(Courtesy of Alfa Laval Marine & Power AB)

Figure 102. Operating Principal of Plate Heat Exchanger

The holes in each corner of the plate form passages for the two fluids, which pass through the narrow passages between the plates. The gaskets on the plates are arranged so that fluids enter through alternate passages - fluid 'A' enters between the odd passages and fluid 'B' enters between the even passages.

Maintenance basically involves cleaning the plates. As with all heat exchangers, irrespective of type, the intervals between cleaning can be determined by monitoring the condition of the cooler. This can be done by using the inlet and outlet temperatures of the fluids or by measuring the differential pressure.

Coolers suffer from three types of deposit:

- Marine growth - algae, weed, barnacles, etc.
- Silt, sand and mud.
- Scale.

Water heaters are mainly susceptible to scaling. Fuel and lubricating oil heaters are susceptible to coking.

Opening up a plate heat exchanger for cleaning.

- Close all inlet and outlet valves.
- Allow plates to cool down.
- Vent and drain the heat exchanger.
- Clean and grease the top and bottom guide bars.
- Grease the removable header roller, if fitted.
- Lubricate the threads of the tightening bolts.
- Measure and record the distance between the two headers (top and bottom and on both sides).
- Release the tightening bolts diagonally, a few threads at a time.
- Once the tightening bolts have been removed, slide the plates to the free end of the guide bar.

Cleaning of heat exchanger

- Clean the fixed header and then each plate individually.
- Plates should be washed with water and a soft brush. A wire brush should not be used. If deposits are heavy a high pressure washer may be required.
- Heavy contamination should be removed by soaking in a chemical bath.
- If plates are removed, their location within the plate stack should be noted so they can be refitted in the correct position.
- Gaskets should be inspected for any signs of damage.

Assembly after cleaning

- Make sure all plates and gaskets are clean.
- Clean rubber inserts on fixed header.
- Push plates into position one at a time.
- Slide removable header into position against the plate stack.
- Fit the tightening bolts and tighten diagonally a few threads at a time.
- Check to ensure the plates are compressed correctly. Do this by checking the measurements between headers to ensure they are the same as before opening up. When assembled correctly the outside of the plates should show a honeycomb pattern.

Plate cooler problems

External leakage

- Defective gaskets.
- High pressure or temperature.
- Insufficient compression of plate.

Internal leakage

- Hole in plate due to corrosion or fatigue crack.

Plate stack

- Corroded plate.
- Fouled or buckled plates.

Increased pressure drop

- Fouled plates.

3.4   Fresh Water Generator

Ocean going vessels do not have the space on board for storing large quantities of fresh water, therefore, much of what is required must be made on board. The majority of fresh water generators are of the flash type. These operate on the principle that the boiling temperature of water varies with pressure. By creating a vacuum water will boil at a lower temperature and the engine jacket cooling water can be used as the heat source. A modern plate type evaporator is shown in Figure 103.

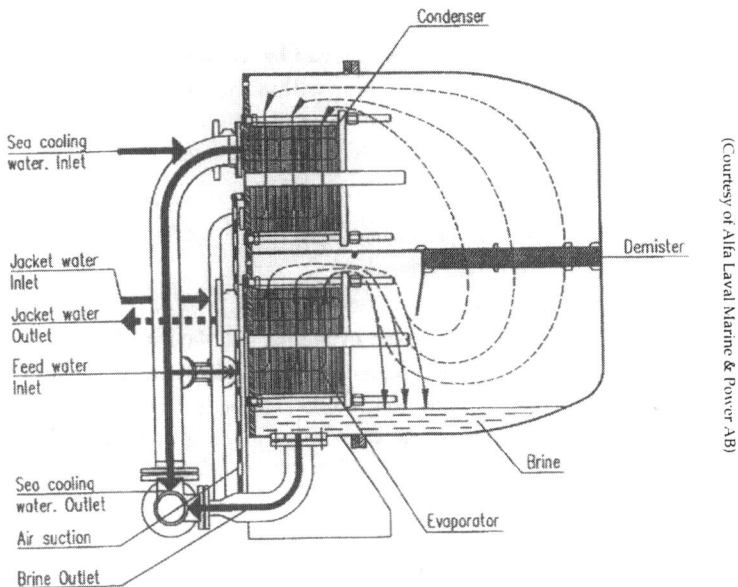

Figure 103. Plate Type Fresh Water Generator

A traditional shell and tube type evaporator is shown diagrammatically in Figure 104.

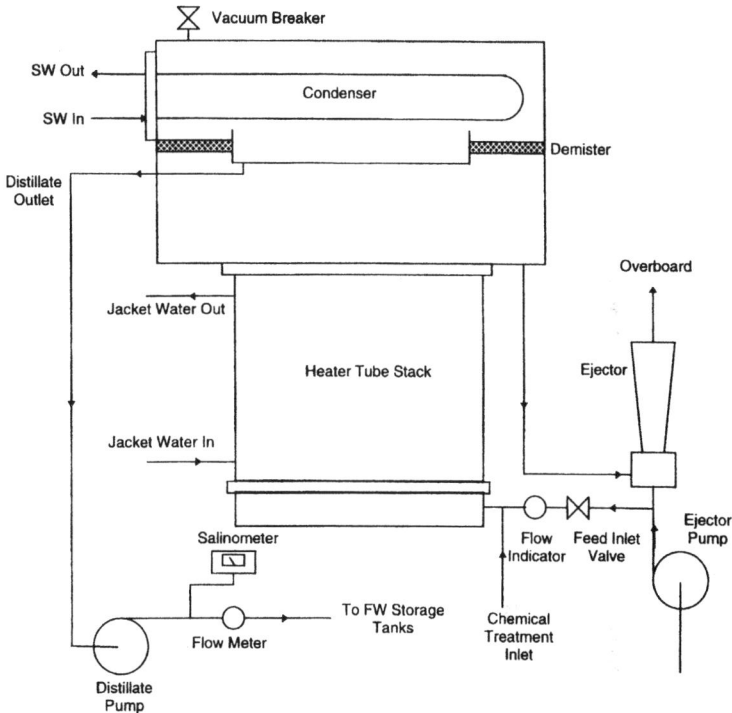

Figure 104. Diagrammatic Arrangement of Fresh Water Generator

In a flash type fresh water generator the engine jacket cooling water is fed to the heat exchanger where it circulates outside the tubes. Sea water feed passes through the inside of the tubes.

The pressure inside the evaporator is maintained at a vacuum by the air ejector and the reduction in pressure results in some of the sea water flashing off into vapour. The remaining brine is taken overboard via the ejector. The majority of fresh water generators, or evaporators, are of the flash type.

The vapour passes upwards through a demister and is condensed by a cooler. The condensate is collected in a tray and fed to the distillate pump. The salinity of the distillate must be monitored and once it falls to a preset value, usually 2ppm, the distillate pump will start to pump the water to the fresh water storage tanks.

151

Operation

- Open up ejector pump system and start the pump.
- Open the inlet and outlet valves for the condenser.
- Open the feed valve and adjust the flow until the flow indicator is at the correct position.
- Allow a vacuum to build up inside the evaporator. This should fall to 700mm Hg.
- When a vacuum is achieved, the heating inlet and outlet valves should be opened to admit jacket cooling water to the heater. This should be done slowly because the heater acts as another jacket water cooler. The jacket water temperature controller may not respond quickly enough if the valves are opened suddenly, resulting in a fall in jacket water temperature.
- Switch on the salinometer. This will be high initially, but should gradually fall as fresh water is produced.
- Set the distillate pump to automatic and open the valves to the fresh water storage tank.
- When the salinity falls to the preset level the distillate pump will start.
- Open up the chemical feed system and set the flowmeter for the correct dosage rate.

The capacity of the fresh water generator is regulated by increasing or decreasing the amount of jacket water to the heater. The evaporation temperature should be between 45-60°C. This will vary depending upon the sea water temperature. When the sea water temperature is low, the evaporation temperature will be lower. It can be raised by reducing the sea water flow to the condenser or adjusting the vacuum, if possible. The temperature rise across the condenser should be about 6°C. When the sea water temperature is high, the sea water to the condenser should be opened up more, if possible.

The fresh water generator should not be used within 50 miles of land or in estuarial water due to harmful bacteria found in the water.

It is important to disinfect the water produced before use. A disinfecting unit is fitted to the domestic fresh water system and may be either ultra violet or an electrolytic salt cell which produces sodium hypochlorite.

Maintenance

The main problem with fresh water generators is scale formation on the heating surfaces. Chemical injection systems are fitted to allow anti-scale chemicals into the feed line to help prevent scale formation. This system is quite often neglected, however, it is important to ensure the flowmeter is working correctly and the correct amount of chemical added to the treatment tank.

Once scale forms on the heating surfaces the evaporator output will fall off rapidly. The best cleaning method is to circulate descalent chemicals through the heater tube stack. A typical system is shown in Figure 105.

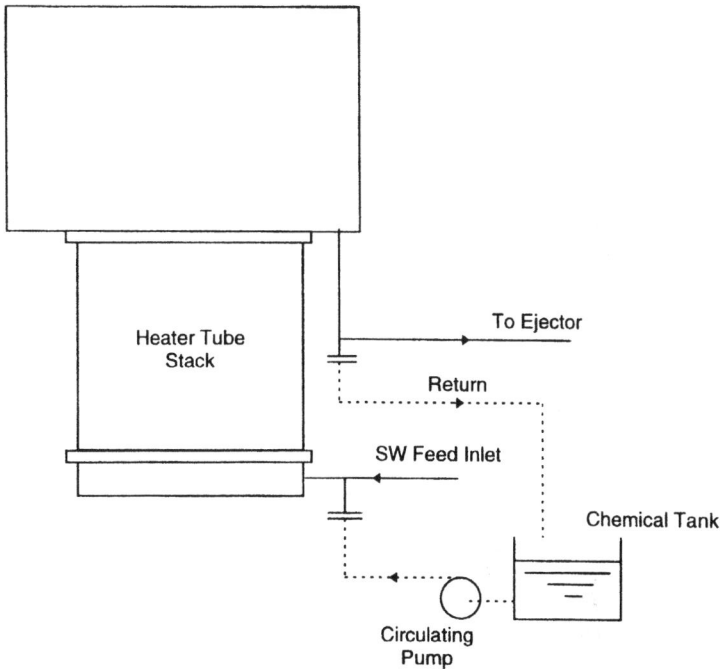

Figure 105. Arrangement for Chemically Cleaning Heater Stack

Chemicals should be circulated for a period of time and the chemical manufacturer's instructions should be referred to. The descaling chemical usually has a built in colour indicator to show when it has been exhausted. The tube stack should then be drained and flushed with fresh water. It is impossible to determine the condition of the tubes without removing the lower cover, therefore, the performance of the unit should be noted after each cleaning session. For plate type evaporators, the plate stack can be removed and submerged in de-scalant.

Occasionally the fresh water generator should be inspected internally. Older models may have a rubberised coating which may start to blister or crack. Any damaged areas should be cut away and the metal underneath thoroughly cleaned of rust and then allowed to dry. Only specially approved coatings should be used for repair and the instructions should be followed carefully as ventilation may also be required. Modern fresh water generators have no coatings and are constructed from corrosion resistant materials,

When reassembling after maintenance, ensure all joints and connections are in good condition and have been tightened properly.

Fresh Water Generator Problems

PROBLEM	CAUSE	ACTION
High Salinity	Capacity too high	Reduce feed
	Evaporation temperature too low	Reduce sea water cooling
	Brine level too high	Ejector nozzle clogged or ejector pump overboard closed
	Condenser tube leaking	Find and plug leak
	Defective or dirty salinometer cell	Replace/clean cell
Low Capacity	Heater tubes fouled with scale	Clean
	Air Leaks (low vacuum)	Find leaks and seal
	Defective ejector (low vacuum)	Inspect and replace
	Low jacket water temperature	Increase temperature
	Defective distillate pump	Check pressure and load

Table 13. Fresh Water Generator Problems

3.5 Refrigerating & Air Conditioning Plant

This section covers small domestic refrigerating systems and air conditioning systems found on board ship. For further information on refrigerating plant the reader is referred to MEP Volume 1, Part 4 - Refrigeration Machinery and Air Conditioning Plant.

Basic Principal of Operation.

Domestic refrigeration and air conditioning systems are usually of the direct expansion type, that is, the entire system is fully flooded with one refrigerant. The system consists of a compressor, condenser, expansion valve and evaporator.

Refrigerant is discharged from the compressor as a superheated vapour. This passes into the condenser where latent and sensible heat is removed. The liquid leaves the condenser undercooled by a few degrees to minimise the amount of refrigerant 'flashing off' before the expansion valve. Any refrigerant flashing off before the expansion valve results in a loss of refrigerating capacity.

The liquid refrigerant then expands across the expansion valve, reducing its temperature prior to entering the evaporator, which is situated in the refrigerated chamber. Here the liquid evaporates absorbing latent heat and reducing the chamber temperature. The expansion valve is adjusted so that the vapour leaves the evaporator slightly superheated, ensuring no liquid returns to the compressor.

Refrigerants and the Montreal Protocol

The desirable properties of a refrigerant are:

- High coefficient of performance (COP) (this is the ratio of cooling effect to the power consumed by the plant).
- Low boiling point.
- Low critical temperature.
- Low condensing pressure.
- High specific enthalpy.
- Low specific volume.
- Good stability (does not react with lubricant or materials in the system).
- Non flammable.
- Non toxic.
- Good leak detection properties.
- Must be readily available.

The refrigerants traditionally used on board ship are Freon 12 and 22, commonly termed R12 and R22 respectively. R12 (CCl_2F_2) is a member of a family of refrigerants known as chlorofluorocarbons (CFCs) which contain chlorine, fluorine and carbon. R22 ($CHClF_2$) is a hydrochlorofluorocarbon (HCFC) as it contains hydrogen, chlorine, fluorine and carbon.

R12 was first introduced in the 1930's and became widely used, but in 1974 it was discovered that CFCs cause ozone depletion. CFCs are very stable and do not break down easily, however, in the stratosphere they are broken down by the action of ultra violet light releasing chlorine atoms. The free chlorine combines with ozone (O_3) to form chlorine monoxide (ClO) and oxygen (O_2). The first clear sign of damage to the Earth's ozone layer over Antarctica was in 1985, and the Montreal Protocol of 1989 laid down criteria for the production and consumption of CFCs and halons, which are used for fire extinguishing.

Initially, under the Montreal Protocol, from 1989 CFC production and consumption was to be frozen at 1986 levels and reduced by 20 per cent from 1993 and then 50 per cent from 1998. Halon production and consumption was to be restricted to 1986 levels as of 1992. Originally HCFCs such as R22 were not included as they have a much lower Ozone Depletion Potential (ODP) than CFCs and it was envisaged that these would be phased out between 2020 and 2040.

Revisions to the Montreal Protocol have laid down stricter controls and HCFCs have been included. The European Community (EC) imposed stricter phase out dates with the phase out of CFCs from January 1, 1995. The EU has banned R22 in new equipment over 150kW as of January 1, 1999.

Refrigerants also contribute towards global warming and are rated according to their Global Warming Potential (GWP). The total effect of the whole refrigeration plant towards global warming is given as a TEWI (Total Equivalent Warming Impact) figure. This takes into account how energy efficient the plant is as well as the GWP of the refrigerant in use.

HCFCs such as R22 are used in over 80 per cent of ships at present and will continue to be available for several years to come, although environmental concerns may result in a reduction in the timescales for phase out. All of the above has led to the development of refrigerants with zero or minimal ODP. R134a has been developed as a replacement for R12 and this has been used in new refrigerated containers. R134a is a hydrofluorocarbon (HFC) and as it has no chlorine, it has zero ODP. R134a can also be used to replace R12 in existing systems, although there is some loss in cooling capacity and R134a is not compatible with mineral lubricating oils, so synthetic oils must be used. The latest replacement for R12, with similar performance, is R407d which is a 15/15/70 blend of R32, 125 and 134a respectively, and it has zero ODP.

R22 can also be used to replace R12 in existing systems but changes to the plant, such as the compressor motor, relief valves, oil separators, filters and instrumentation, are required. The condenser pressures are higher, therefore, the system will need pressure testing.

A direct replacement for R22, which has similar thermophysical properties and can be used in existing systems, is more difficult to find. At present R407c, which is a 23/25/52 blend of R32, R125 and R134a respectively, seems suitable for most but not all systems. R410a has also been developed as a replacement for R22 for new systems and is a 50/50 blend of R125 and R32.

Stopping and starting a Compressor

The following is a list of steps for a typical refrigerating plant. Some systems may differ slightly, therefore, the compressor manufacturer's instructions should always be followed.

Stopping and starting refrigerating plant compressors is usually for the routine change over of compressors.

Stopping the running compressor

- To ensure uniformity always stop the running machine by pumping the refrigerant in the system to the condenser.
- Close the condenser outlet valve. The suction pressure will fall and the compressor will eventually stop on low pressure.
- Set the compressor control to manual at the control box.
- Close the compressor suction and discharge valves and condenser valves.
- Close the condenser cooling water valves.
- Ensure the crankcase heater is on.

Starting the standby compressor

- Check the crankcase oil level is satisfactory.
- Open the compressor discharge valve and condenser valves.
- Open up the cooling water to the condenser.
- Crack open the compressor suction valve.
- Set the standby compressor to 'automatic' on the control box.
- Once the compressor starts and the suction pressure begins to fall, gradually open up the suction valve until fully open. This ensures no liquid can enter the compressor, which could cause damage.
- Check the oil pressure.

Note: The oil pressure gauge reads oil pressure plus suction pressure, so the actual oil pressure is the oil gauge reading minus the suction gauge reading.

- Check the compressor oil level and liquid sightglass while the system settles down. If the standby compressor was shut down, as above, there should be a satisfactory charge in the system.

Shutting the system down by pumping the refrigerant to the condenser ensures the compressors are shut down in a similar condition. It also ensures there is always a spare charge of refrigerant ready to use if there is a problem.

The operation is fairly similar for air conditioning systems. If the compressor needs to be stopped for a short period, overnight for instance, the compressor should be stopped and the suction valve closed. When restarting, the suction valve should be cracked open prior to start-up and then once the compressor has started and the suction pressure starts to fall, the valve should be gradually opened up. This ensures no liquid enters the compressor.

When stopping the plant for a longer period, such as a seasonal shut down, the same procedures should be followed as for starting and stopping refrigerating compressors.

Leak Detection

This should be a routine operation carried out after any pipe connections have been broken or the compressor opened up. The complete plant should be checked including pipe connections, valve glands, pressure gauges and around the compressor, particularly the shaft seal. As the refrigerant is heavier than air it will sink, therefore, it is important to check the bottom of the connections. Oil may also be present where there is a leak. Leaks can be detected with either soap solution, a halide torch or an electronic leak detector.

Soap solution should be brushed on lightly and observed for any bubble formation.

A halide torch is a small propane burner with a copper element which has a tube attached. The free end of the tube is held close to the possible leak source. Any leaking refrigerant will be drawn up the tube and burnt. When halogen refrigerants are burnt in the presence of copper a bright green flame is produced.

Note: R12 burns to produce phosgene gas, which is poisonous. A halide torch will not detect non halogen refrigerants.

Electronic detectors are the most sensitive and produce a high pitched bleeping sound when refrigerant is detected.

Service Stop Valves

Service stop valves are fitted to compressors and have a service connection used for charging or connecting to a pressure gauge. The valve spindle has a square end which is operated by a special spanner and a cap covers the spindle when it is not in use. The service connection is usually 1/4 inch BSP and should also be capped when not in use. The valve can be backseated and the usual operating positions are shown in Figure 106.

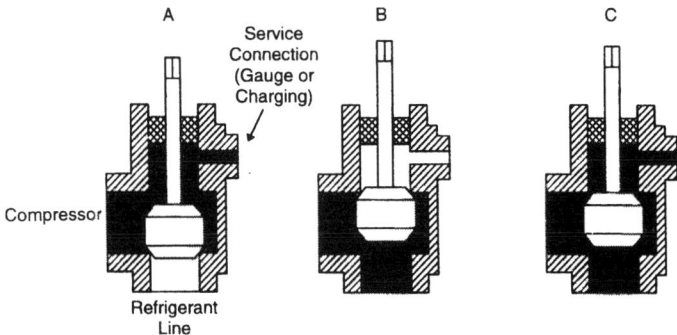

Figure 106. Compressor service stop valve

Position A -   Valve is in the closed position and the spindle is turned fully clockwise. The refrigerant line is closed off and the compressor is common with the service connection. This is the position used during charging.

Position B -   The valve is backseated and the spindle is turned fully anti-clockwise. The refrigerant line is common with the compressor and the service connection is closed off. This is the position during normal operation when the service connection is only used for charging.

Position C -   The valve backseat is cracked open. The spindle is turned fully anti-clockwise and then turned clockwise by one quarter to one half turn. The refrigerant line, compressor and service connection are common and this position is used where a pressure gauge is normally connected to the service connection. The valve should be adjusted to prevent the gauge needle fluttering excessively.

Refrigerant Charging

Refrigerant should be added using a gauge manifold, as shown in Figure 107. This has low pressure (LP) and high pressure (HP) gauges and valves which are usually coloured blue and red respectively. Hoses should also be coloured similarly.

Refrigerant is usually added to the low pressure side of the system, as smaller amounts can be added causing less risk of overcharging the system.

Note: Gloves and goggles must be worn whenever charging the system with refrigerant.

-    Connect a hose between position 'B' and the compressor suction service stop valve. Ensure the valve is fully backseated before connecting the hose.

-    Connect a hose between position 'A' and the refrigerant cylinder. Fit a vacuum pump into this line using a 'T' piece.

-    Close the manifold HP valve and open the LP valve.

-    Start the vacuum pump to remove air from the charging hoses. Stop the pump when this is complete.

-    Close the manifold LP valve.

-    Close the condenser outlet valve and allow the compressor to pump all refrigerant to the condenser. The compressor will cut out on low pressure.

-    Check and note the liquid level in the condenser if a sightglass has been fitted.

-    Close the compressor suction service stop valve.

- Open the vapour valve on the refrigerant cylinder. Large cylinders usually have two valves, vapour and liquid, coloured blue and red respectively. Small cylinders may only have one valve and the cylinder must be turned upside down for vapour.
- Crack open the LP valve on the gauge manifold. The compressor suction pressure will rise until the compressor starts.
- The compressor is now taking refrigerant vapour from the cylinder and pumping it to the condenser.
- Allow the level in the condenser to rise slightly.
- Close the LP gauge valve and cylinder valve.
- Open the condenser outlet valve.
- Slowly open the compressor suction valve. Once the compressor is running open it until fully backseated.

Once the system has settled down check the liquid line sightglass for bubbles and check room temperatures. The cylinder should be weighed before and after charging and a note made of the quantity added. If there is no condenser sightglass then only add a small quantity of refrigerant at a time as it is easy to overcharge the system.

Always ensure there is sufficient refrigerant on board, because a leak in a system can quickly lead to a complete loss of refrigerant charge.

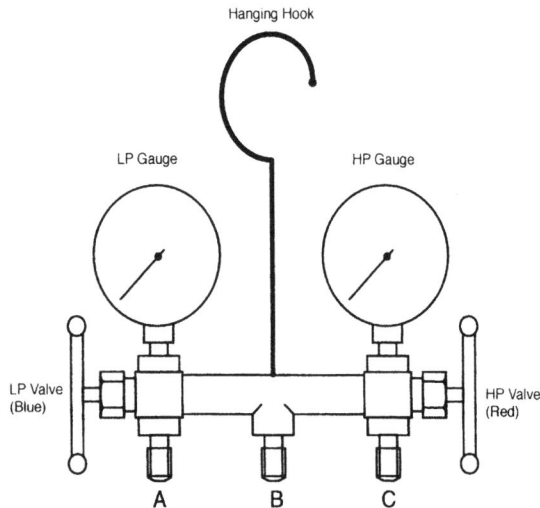

Figure 107. Refrigerant Gauge Manifold

Problems and Maintenance

When carrying out any adjustments on refrigeration or air conditioning plant always keep a note of the work carried out. Make small adjustments at a time, allowing the plant to settle down. Monitor any changes before making further adjustments.

Table 14 covers some of the problems that may occur, however, the system needs to be viewed as a whole. There may be several symptoms caused by one fault. Problems are usually due to a lack of refrigerant which can cause several different problems to occur.

The usual indication of this is the presence of bubbles in the liquid line sightglass. The compressor may also be running continually just to maintain room temperature. Under normal conditions the sightglass should be clear, though there may be a few bubbles when the compressor first starts. If bubbles are present, check the liquid level in the condenser, if a sightglass has been fitted. Liquid should be just visible in the condenser sightglass under normal conditions. If there is too much liquid the lower tubes in the condenser may be covered which will reduce the effectiveness of the condenser.

Bubbles in the liquid line sightglass may be due to air which can enter during maintenance. Air can also enter if the compressor is drawing a vacuum. The LP cut-out should be set above zero to prevent this. If air is present the compressor discharge pressure will increase. Air in the system can be checked by pumping the refrigerant down to the condenser, until the compressor stops, then closing the compressor discharge valve. If there is no air in the system the cooling water temperature should correspond to the condensing temperature on the HP pressure gauge. If the gauge reading is higher, air may be present in the system. Any air should be bled off the system at the condenser.

Moisture should not be present in a system, but can enter during maintenance. Moisture in the system can freeze at the expansion valve, blocking the line. Refrigerants purchased from a reputable supplier will be dry, but this may not always be the case with unknown suppliers. The liquid line sightglass usually has a moisture indicator which shows wet and dry, and filter/driers are fitted in the liquid line. Spares should be stored with the end caps on and changed according to the manufacturer's recommendations.

Problem	Cause	Action
High compressor discharge pressure	Air in system	Bleed off air at condenser
	Insufficient condenser cooling:	
	Condenser fouled	Clean tubes
	High coolant temperature	Reduce coolant temp. If not possible then monitor pressure
	Insufficient coolant flow	Adjust flow
	Overcharge of refrigerant	Reduce charge
	Discharge valve closed in	Check valve is fully open
	Pressure gauge defective	Check gauge operation
High compressor suction pressure	Expansion valve open	Check fridge doors closed
		Check expansion valve
	Suction or discharge valves leaking	Overhaul valves
	Capacity control inoperative	Check operation
	Pressure gauge defective	Check gauge operation
	Safety valve leaking	Overhaul valve
Low compressor discharge pressure	Overcooling in condenser:	
	Low coolant temperature	Reduce flow of coolant
	High coolant flow	Reduce flow of coolant
	Leaking discharge valves	Overhaul valves
	Worn piston rings	Overhaul
	Liquid returning to compressor	Check expansion valve
Low compressor suction pressure	Shortage of refrigerant in system	Charge system and leak test
	Suction filter of expansion valve filters choked	Clean filters
	Expansion valve shut in	Adjust valve
	Suction valve not fully open	Check valve is fully open
	Suction gauge defective	Check gauge operation
Compressor icing on suction side	Excessive charge of refrigerant	Reduce charge
Compressor runs continually	Shortage of refrigerant in system	Charge system and leak test
	Leaking compressor valves	Overhaul valves
Noisy compressor	Liquid refrigerant returning:	
	Excessive charge of refrigerant	Reduce charge
	Expansion valve wide open	Adjust valve
	Worn compressor	Overhaul
Short cycling (HP cut out)	Low setting of cut out	Check and reset
	See high compressor discharge	
Short cycling (LP cut out)	LP cut out set too high	Check and reset
	Suction filters or expansion valve filters choked	Clean filters
	Faulty expansion valve	Check valve
	Evaporator iced up	Defrost
Low oil pressure	Low crankcase oil level	Top up oil level
	Oil filters dirty	Clean or replace filters
	Pressure regulator problem	Adjust or overhaul
	Defective pressure gauge	Check gauge operation
	Worn bearings	Overhaul

Table 14. Refrigeration Plant Problems

Routine maintenance involves inspecting the liquid line sightglass for bubbles and moisture, checking the compressor oil level, leak testing, checking compressor drive belts (if fitted), inspecting evaporators for icing, and testing compressor cut outs. Other maintenance involves changing the filter/drier, cleaning the compressor suction filter and expansion valve filters, and compressor oil filter cleaning and oil change, all of which should be carried out at intervals as specified by the compressor manufacturer.

Oil can be topped up without stopping the compressor by using a hand pump to force oil into the crankcase.

Another task that may have to be carried out is replacement or alteration of piping, particularly if a longer or shorter filter/drier needs to be fitted. Tubing in small systems is mainly copper which has been annealed so it can be bent and flared easily. Spare coils of tubing should be kept on board and the ends should be capped to prevent dirt and moisture ingress. When bending tubing use a bending spring or tube bender. A special cutter should be used for cutting tubing ensuring that burred ends are cleaned up. If tubing has to be sawn, ensure no filings are left in the tube.

Most fittings have flared connections. A special tube flaring tool should be used for making flares, as shown in Figure 108.

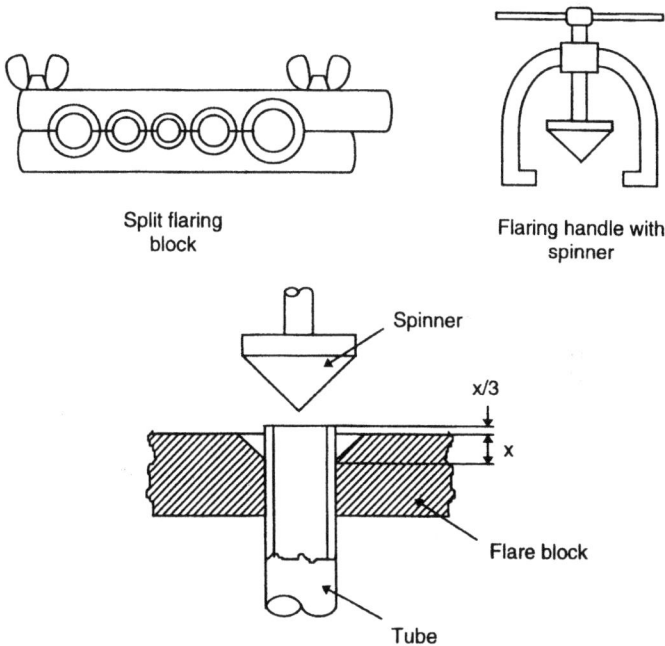

Split flaring
block

Flaring handle with
spinner

Figure 108. Tube Flaring Tool

Ensure the end of the tube is squared off and smooth, then fit the tube into the split flaring block, leaving the tube slightly proud of the block, as shown. Make sure the nut is fitted onto the pipe as it will not be possible to fit this once the flare has been formed. Lubricate the spinner face with refrigerant oil and tighten the spinner onto the pipe to form the flare.

4 Boilers & Boiler Water Treatment

4.1 Auxiliary Boilers and Burners

Auxiliary boilers on motorships vary from small units which supply low pressure steam for domestic heating purposes to larger high pressure boilers capable of supplying superheated steam for cargo pumps on tankers. Most modern marine auxiliary boilers operate on residual fuel and have steam assisted atomisation.

The intermittent operation of auxiliary boilers and their relatively low pressure may result in them not getting the attention and maintenance they deserve, particularly the boiler feed water.

The auxiliary boiler is generally used in port and shut down at sea, though some boilers may use the steam drum in conjunction with the exhaust gas boiler. Prior to arrival in port the boiler will need to be flashed to get it ready for putting on line.

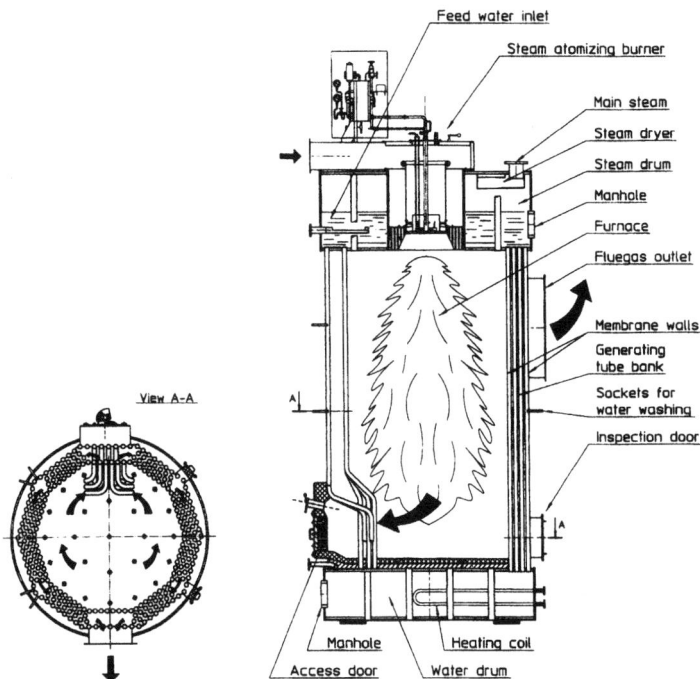

Figure 109. Aalborg AQ18 Two Drum Boiler

Starting Auxiliary Package Boiler

The following will not apply to all boilers and some additional steps may be required, depending upon the system. Modern boilers have automated firing and purge periods are set by timers within the control cabinet.

- Ensure the air vent on top of the boiler is open and that there is no pressure in the boiler. Check the steam stop valves are shut.

- Open the feed water valve and fill the boiler with water until it is just showing in the gauge glasses. Then shut the feed water off. Ensure the water level is above the extra low water level alarm, otherwise, the burner will not start. If water is filled to the normal level, it will rise further as the water is heated up and may end up too high when the boiler reaches its operating temperature.

- Circulate fuel oil until it is at the correct temperature. If steam atomisation is used this may need to be changed over to the air supply if steam is not available. If steam or air atomisation is not fitted, the boiler may need to be started on distillate fuel.

- Start the boiler on its automatic cycle. The burner fan will start to purge the combustion chamber and the burner oil pump will circulate the hot oil. After a pre-set purge time the burner will ignite.

- Check the combustion chamber sight glass to ensure the burner has lit and the flame is satisfactory.

- Keep a close check on the water level as the steam pressure increases.

- Once steam comes from the air vent it can be closed.

- When the normal working pressure is reached, blow down the gauge glasses and float chambers for the level alarms.

- Change over to steam atomisation, if fitted.

Shutting Down the Boiler

- If the boiler is to be stopped for an extended period or opened up for maintenance, change the fuel system to distillate fuel.

- Stop the boiler automatic cycle.

- Close the steam stop valves.

- Ensure fuel is circulated if left on residual fuel operation.

- Close the feed valves.

- When the boiler pressure has reduced to just above atmospheric, open the vent cock to prevent a vacuum forming within the boiler.

Blowing Down the Boiler

The reasons for fitting scum and blowby valves are to:

- Reduce the density of the boiler water by reducing the water level and refilling it with distilled water. (See 4.4 Feedwater Treatment).

- Empty the boiler prior to entry.

- Reduce the water level in an emergency.

The scum valve is used to remove floating dirt, foam and oil from the surface of the water.

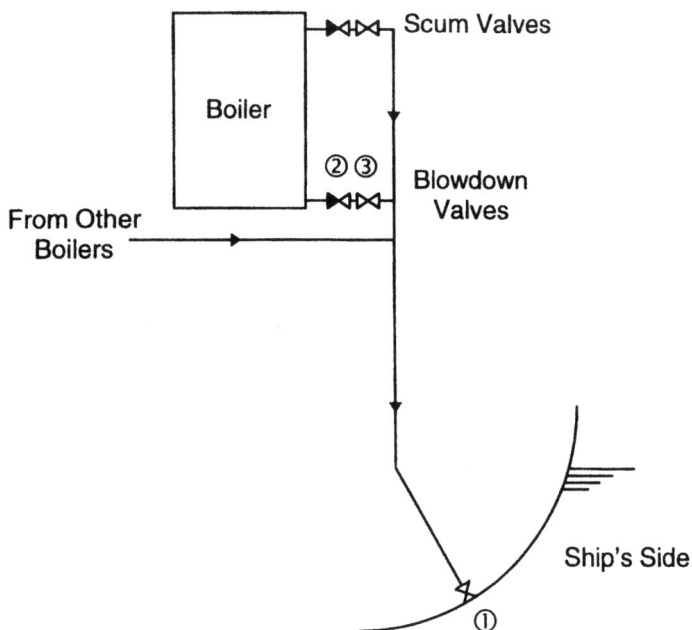

Figure 110. Boiler Blowdown and Scum System

Open valves in order (1), (2) and (3), and close in reverse. The blowdown valve adjacent to the boiler (2) should be fully open to prevent cutting of the valve seating faces. The rate of blowdown is controlled by the second blowdown valve (3). Never leave the boiler unattended when blowing down and ensure all valves are fully closed after blowing down. A hot drain pipe indicates a leaking blowdown valve.

Draining the Boiler

The boiler should always be allowed to cool before draining otherwise deposits and sludge can become baked onto tubes and other hot surfaces. Open the vent valve when the boiler pressure has fallen to just above atmospheric, to prevent a vacuum forming within the boiler.

4.1.1   Boiler Burners

Modern boiler burners are designed for operation with residual fuel. Fuel is circulated to the burner continuously and may have an electric heater with electric trace heating on the fuel piping. Burners usually have steam assisted atomisation, with air operation during start up periods. Pilot burners, operated on distillate fuel may also be used to ignite the main burner. A steam jet type nozzle is shown in Figure 111.

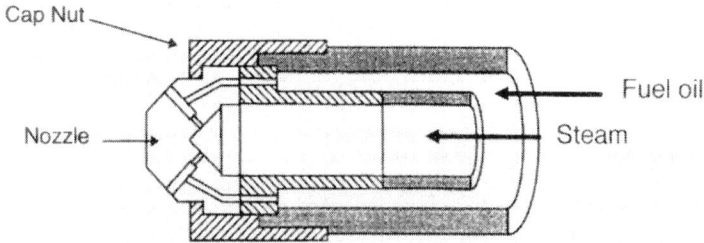

Figure 111. Steam Jet Type Burner Nozzle

During start up of the burner there should be a purge period where the fan blows air through the furnace. This is built into the automatic programme for starting the burner. If the burner has to be operated manually the fan should be run for a few minutes to purge the furnace prior to ignition. If the burner is failing to fire, unburnt fuel will remain in the furnace. If this is hot, fuel vapour will build up inside the furnace. If it is not purged before the next attempt at firing, there is a risk of these vapours igniting and causing an explosion.

Burner Maintenance.

Burner maintenance mainly involves cleaning burner nozzle(s), swirl plate, igniter electrodes and filters. The boiler manufacturer's guidelines should be referred to.

Nozzles should be cleaned at regular intervals, to remove deposits that build up on the nozzle tip. Brass wire brushes should be used for this as steel brushes can damage the nozzle holes. Nozzles may also have filters within their assembly which should be washed out and blown through with compressed air.

Igniter electrodes should be checked for condition, correct gap and distance from burner. Ceramic insulators can be coated with soot deposit, which can affect igniter performance. Any soot should be cleaned off. The condition and security of electrical leads should also be checked.

PROBLEM	CAUSE	ACTION
Burner does not light	Fuel pressure too low	Raise pressure, check filters in suction and burner. Fuel pump safety valve may need adjustment
	Fuel temperature too low	Increase temperature, check thermostat
	Dirty or defective magic eye	Clean photocell
	Burner electrodes not sparking	Check through viewing port for spark. Check condition of electrodes and gap. Problem with ignition transformer or relay
	Control panel in alarm	Rectify fault and reset
Poor atomisation	Clogged burner tip	Clean burner
	Low fuel or air pressure	Check filters

Table 15. Burner Problems

4.2   *Exhaust Gas Boilers*

Approximately 28 per cent of the energy in the fuel supplied to the engine disappears up the funnel. Roughly half of this can be regained by using the energy of the exhaust gas to produce steam in an exhaust gas boiler. The size of the engine installation determines the size of the waste heat recovery system and large engine installations produce sufficient energy to produce superheated steam.

Exhaust gas boiler systems vary and it is not possible to cover them all here. Figure 112 shows a typical system for tank heating and domestic hot water.

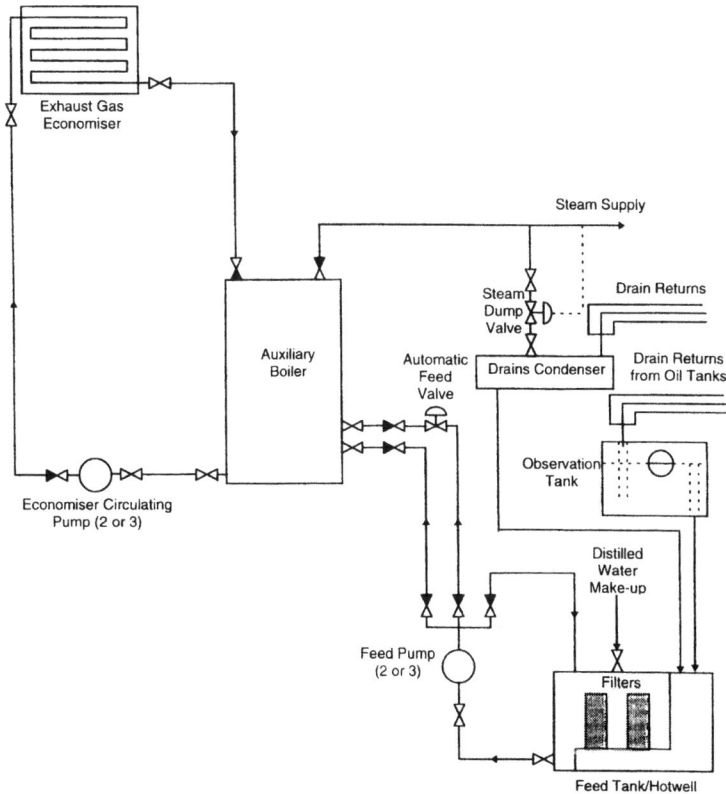

Figure 112. Typical Steam System with Exhaust Gas Economiser

Figure 113 shows a system for producing superheated steam. Many of the practices for auxiliary package boilers, such as blowing down, draining and routine inspections, apply to exhaust gas boilers.

169

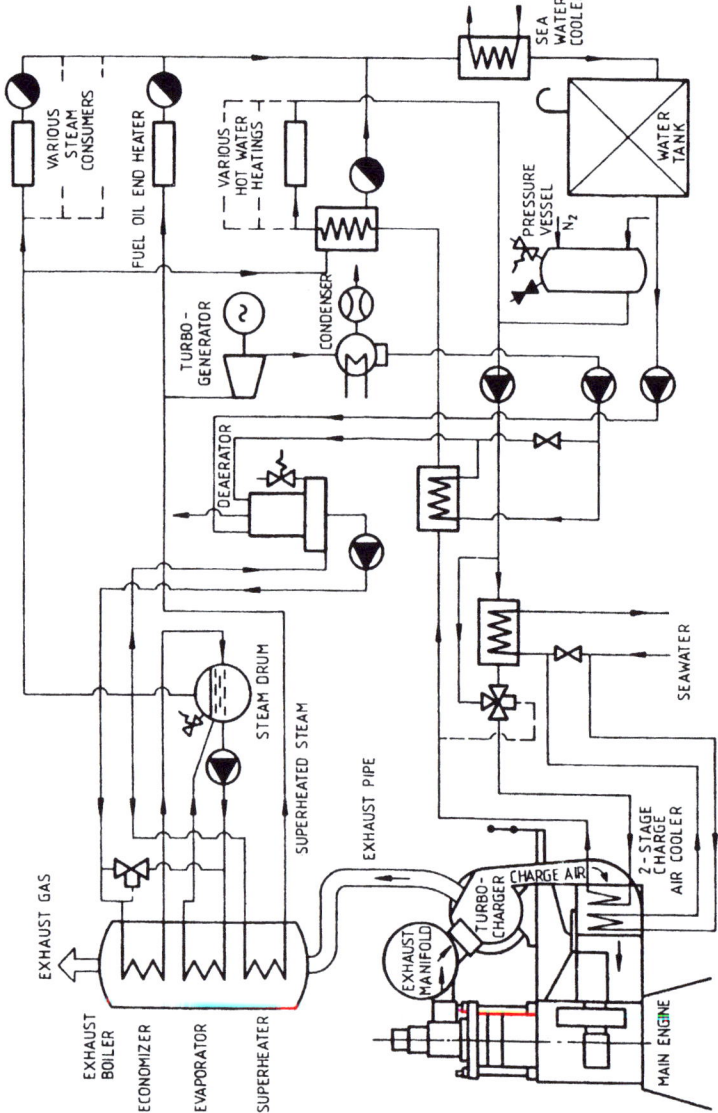

(Courtesy of Wärtsilä NSD)

Figure 113. Waste Heat Recovery System for Superheated Steam

Starting the Exhaust Gas Boiler

The following should be used as a guide and will not apply to all boilers. Some additional steps may be required, depending on the system.

- Check there is no pressure in the boiler then open the boiler air vent.
- Open the feed valve and gradually fill the boiler until water appears in the gauge glasses. The feed valve can then be closed. The reason for not filling the boiler to the normal level is that the water level will rise as it heats up and may end up too high when normal working temperature is reached.
- Open the valves to the circulation pump and start the pump. Check the boiler water level after starting the pump.
- Gradually admit exhaust gas to the boiler to allow the boiler to warm up slowly. Rapid heating will increase the stress on components due to different expansion rates. Keep a regular check on the boiler water level.
- Close the vent cock when steam issues from it.
- Slowly open the main steam stop valve once the steam pressure rises.
- Open up the feed valve and ensure it is set to automatic operation.

Shutting Down the Boiler

- Before shutting down the boiler, operate the soot blowers.
- Bypass the exhaust gas.
- Close the steam stop valves.
- Maintain the water circulation for at least 12 hours after stopping the main engine.
- When the boiler pressure has fallen to just above atmospheric pressure, open the air vent to prevent a vacuum forming inside the boiler.
- Close the feed valves.

4.2.1   Uptake Fires

Engine exhaust gas contains particulates which consist of partially burnt fuel and/or lubricating oil and ash from the fuel and lubricating oil. These particulates can form deposits on boiler tubes. The situation can be made worse by prolonged low load operation of the engine, which reduces exhaust gas velocity and may lead to soot deposition, burning poor quality fuel, or poor combustion due to defective injection equipment or inefficient fuel treatment. Incomplete combustion leads to the formation of sticky deposits.

If the boiler runs dry any deposits on the tubes will quickly be heated to exhaust gas temperature. The excess oxygen in the exhaust gas will oxidise the deposits which may cause the deposit temperature to rise above self ignition temperature. Soot deposits may be ignited by glowing carbon particulates in the exhaust gas. The ignition temperature of the soot is usually less than 400°C, however, if deposits are sticky it could fall to less than 200°C. It is possible for soot fires to occur after the engine has been shut down, therefore, it is important to maintain water circulation after shut down. If dark smoke is coming from the funnel or there is a large rise in exhaust gas temperature after the boiler, it could indicate the presence of a fire.

Fires can be avoided by carrying out regular soot blowing and maintaining full water circulation. Likewise, if a rise in the exhaust gas temperature differential across the boiler occurs, it could indicate fouling of the tubes. If a fire does occur the engine should be stopped immediately and the turbocharger air intakes covered to starve the fire of air. The soot blowers should not be used as this could fan the fire. Check to ensure full water circulation is maintained. A small fire may burn itself out as the heat will be conducted away by the circulating water. If a water washing system is fitted it could extinguish a small fire. If there is a tube leak, the circulation should be stopped and the boiler drained.

If the circulation stops it causes a risk of tube failure which can cause a hydrogen fire. Without circulation the tubes will rapidly increase in temperature and may eventually rupture if under pressure. The resulting water vapour dissociates releasing hydrogen which will burn reaching temperatures as high as 2000°C. The heat sustains the reaction causing the iron to oxidise which can melt the complete tube stack. In case of a high temperature fire the engine should be stopped immediately. The boiler circulation should be stopped and the boiler drained. Boundary cooling should then be carried out until the fire is extinguished.

Any fire should be dealt with quickly to prevent a small fire becoming a large one which could destroy the boiler.

Soot Blowers

Soot blowing should be carried out on a regular basis to ensure soot and ash deposits do not build up on the tubes. Soot blowing should also be carried out after water washing the main engine turbochargers (turbine side) and prior to shutting down the boiler. More frequent blowing should be done when the engine is operating at low load or if the fuel is known to have a high ash content.

Fire Side Water Washing

The combustion of residual fuel results in the formation of sodium and vanadium slags. These slags gradually build up on the tubes and may eventually bridge the gap between the tubes. Soot blowing keeps these deposits at low levels, however, it does not reach all areas of the boiler.

These slags are soluble in water and can be easily removed with hot water washing. Some boilers have water washing connections, although if they do not have these connections the inspection doors should be removed and the tubes washed with hot water, preferably a high pressure washer. This also applies to oil fired boilers.

4.3   Boiler Maintenance & Inspection

Note: Any welded repairs should not be undertaken without the approval of the appropriate Classification Society.

IMPORTANT: *Before opening up the boiler it must be vented to prevent the formation of a vacuum within the boiler. Serious injury can occur if the boiler is opened with a vacuum inside.*

- Open the boiler vent valve when the pressure is just above atmospheric to prevent a vacuum forming inside the boiler.

- Allow the boiler to cool and then drain.

- Open all manhole, handhole and mudhole doors and allow the boiler to ventilate.

- Isolate all feed and steam stop valves as well as the burner controls.

- Ensure another person is standing by the boiler when someone is inside.

- Plug the blowdown pipe to prevent dirt and scale blocking it. Post a note in a prominent position indicating the blowdown is plugged.

- Thoroughly clean the boiler internally. Remove any scale and clean the tubes internally and externally prior to inspection. Gloves, a dust mask and goggles should be worn when working inside the boiler.

- Inspect the boiler externally for any signs of leakage. Examine the condition of insulating material and replace as necessary. If insulating material contains asbestos it should be wetted before removal to prevent dust formation. Gloves, a dust mask and goggles should be worn when handling lagging. Check all boiler supports are satisfactory.

- Inspect all mounting flanges and studs for security and any signs of corrosion, removing lagging for access if necessary. Wastage can occur where there is a leakage of steam or water.

- Inspect the boiler internally for any signs of cracking, pitting or other corrosion, particularly in way of openings in the shell, around welded connections, and where tubes are welded to the tube plate. Also check for any scale deposits and signs of oil. Check for any signs of blockage, particularly the blowdown pipe. Also inspect the scum and feed distribution pipes.

- Tubes should be checked for signs of distortion which may be caused by overheating - a torch shone across the tubes can be used to show up any deformation. Smoke tubes should be brushed through and water tubes washed externally with a high pressure washer to remove combustion slag.

- Joint faces of manhole, handhole and mudhole doors should be checked to ensure their sealing faces are in good condition.

   - Ensure the door is centralised in its opening.
   - Check the joint faces are in good condition.
   - Ensure clearance is 1.5mm all around (see Figure 114).
   - The nut should be re-tightened when the boiler is at working pressure.

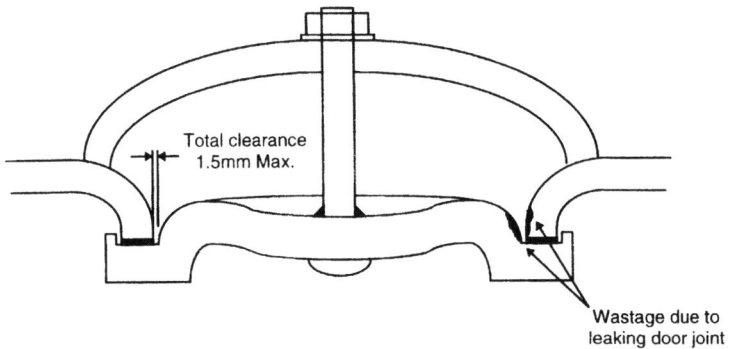

Total clearance 1.5mm Max.

Wastage due to leaking door joint

Figure 114. Boiler Manhole Door

- Inspect the refractory lining for cracks or missing refractory, which should be repaired. The sodium and vanadium compounds in the fuel can result in a loss of refractory material in way of the flame, therefore, check that any loss is not excessive.

- Inspect the combustion chamber and uptake for any signs of combustion deposits.

- Check all valves and gauges are not leaking and that all cocks are clearly marked to show when open or closed.

- Overhaul and inspect the safety valves:
   - Ensure there are no ridges on the valve lid and seat.
   - Check the valve body drain is clear.
   - Check the easing gear is free to operate. Check at zero pressure and inspect all wires and pulleys.

- Check all parts are free to move in the valve body.
- Check the valve spindle for corrosion. Ring test the spindle and test for distortion by running between centres in a lathe.
- Check the spring for corrosion and free length.
- Check the valve chest for corrosion and wastage. Also check the condition of the flange face.
- Ensure the pipe clamps are secure so that the drain pipe is not exerting any force on the valve body.
- Check the operation of boiler controls and alarms:
  - Low fuel temperature.
  - Low fuel pressure - There may be two settings, 'alarm' and 'shutdown'.
  - Low air pressure.
  - Atomising steam pressure low.
  - High steam pressure.
  - Low steam pressure.
  - Low water level - Close the feed valve and let the level fall. Check the alarm sounds.
  - Extra low water - Reduce the feed water level further until the second low level alarm activates and the burner locks out.
  - High water - Manually operate the feed valve until the alarm sounds.
  - Flame failure - During the firing sequence remove and cover the photocell. Test the alarm to ensure it sounds.

Maintenance of the steam system mainly involves checking the valve glands. These should be nipped up as leakage will cause deposits around the gland as well as soaking the lagging. *Glands should never be packed while under pressure.* If valves have to be removed from a section of pipe and the pipe can be pressurised, it should be blanked off. Lines should be properly vented and drained before any work is undertaken.

Feed tanks or hotwells will require draining and cleaning, usually during drydock. Filter towelling should be cleaned and inspected regularly and should be renewed if damaged.

Setting Safety Valves

- Ensure the boiler pressure gauges have been calibrated.
- Gag the valve that is not being set.
- Remove the split compression ring(s) from the valve being set and screw down the compression nut slightly.

- Operate the boiler manually and gradually raise the steam pressure until it reaches lift pressure.
- Slacken back the compression nut until the valve lifts.
- Screw up the compression nut until the valve just seats with a light tap on the spindle.
- Recheck the lift pressure and adjust the valve if necessary.
- Measure the distance for the compression rings and fit the correct thickness of rings.
- Recheck the operation of the safety valve.
- Repeat the process for the second valve.

The setting of safety valves for exhaust gas boilers after a survey must be tested by the Chief Engineer. A signed statement must be sent to the Classification Society, stating the date the valves were tested and the pressure at which they lifted.

Blowing Down Gauge Glasses

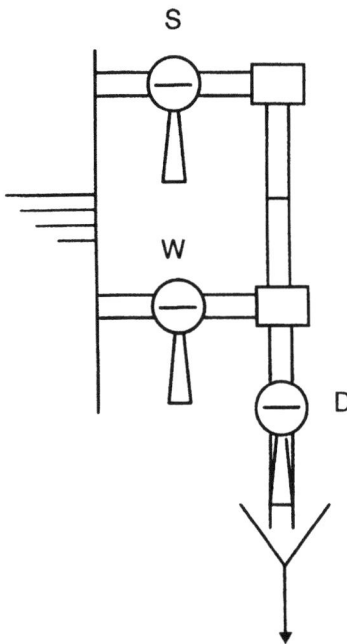

Figure 115. Blowing Down Gauge Glass

There must be two independent means of determining the water level, one of which must be a direct reading gauge glass. The other is usually a remote reading gauge in the control room which should regularly be compared with the boiler gauge glasses.

Close off cocks 'S' and 'W' and open drain 'D'.

Open cock 'W' and blow through. Close cock.

Open cock 'S' and blow through. Close drain 'D'.

Open cock 'W' and check water level rises in glass.

The glasses should be kept clean. The glands should be nipped up at the first sign of leakage as the chemicals in the boiler water will leave a deposit on the glass. The glasses should have protective glass shields around them and be properly illuminated.

Note: Gauge glass cock handles should be pointing down when in their normal working positions.

4.4 Feed Water Treatment

Boiler feed water needs to be treated to prevent corrosion and scale formation within the boiler and feed system. The boiler water needs more than chemicals to keep it in good condition. Regular blowdown and supply of distilled water are also essential.

The purpose of feed water treatment is to:

- Remove the scale forming salts.
- Make the water alkaline.
- Keep dissolved solids to a minimum.
- Maintain the boiler water density by blowdown.
- Provide a reserve of treatment chemicals.

Feed water should always be distilled water and even though pure distilled water contains no scale forming salts it is naturally acidic. Water obtained from ashore may have a high level of salts. Distilled water produced on board may have high salinity if the fresh water generator is not functioning properly. This may occur during rough weather or during start up. The salinity meter may be faulty or the alarm may have been set incorrectly. Condenser leakage can lead to high feed water salinity.

Salts in feed water fall into three categories:

- Those that form scale on heating surfaces.
- Those that cause corrosion.
- Those that remain in solution, increasing the density of the water.

Salts that may occur in boiler water if it is contaminated are as follows:

Sodium Chloride: This should not be found in the boiler under normal conditions. It does not form scale but will cause corrosion and can only be removed by blowdown.

Magnesium Chloride: Does not form scale or corrosion under normal conditions and is removed by chemical treatment with sodium hydroxide.

Magnesium Sulphate: Does not form scale or corrosion under normal conditions and is removed by chemical treatment with sodium hydroxide.

Calcium Sulphate: This is the main scale forming salt in a boiler. The solubility decreases with the increasing temperature, therefore, more scale will be formed at higher temperatures. The scale is hard and difficult to remove, however, it can be removed by chemical treatment with tri-sodium phosphate.

Calcium and Magnesium Bicarbonate: These decompose in the boiler to form their respective carbonates. These carbonates take the form of a soft scale (which can be easily removed), water and carbon dioxide. Carbon dioxide passes over into the condensate with the steam and can cause corrosion in the feed system.

From all of the above it can be seen that the main problems occurring in auxiliary boilers are scale, corrosion and sludge.

Scale

In low pressure boilers the scale forming salts are carbonates and sulphates of magnesium and calcium. Scale forming on heating surfaces reduces the heat transfer and therefore the efficiency. If scale formation is excessive tubes may overheat and split. Chemicals applied to the feed water convert the calcium and magnesium salts to sodium salts which can be removed by blowdown. Chlorides should be kept to less than 200ppm.

Corrosion

Corrosion is a complex process which involves the flow of electrons and the corrosion of a steel plate is shown in Figure 116.

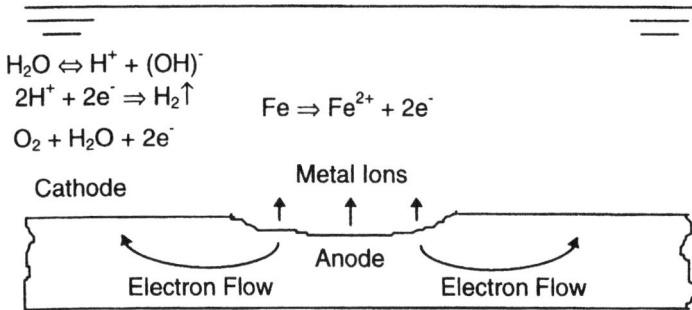

$$H_2O \Leftrightarrow H^+ + (OH)^-$$
$$2H^+ + 2e^- \Rightarrow H_2\uparrow$$
$$Fe \Rightarrow Fe^{2+} + 2e^-$$
$$O_2 + H_2O + 2e^-$$

Cathode   Metal Ions

Anode

Electron Flow   Electron Flow

Figure 116. Corrosion of Steel Plate

Two adjacent areas can exist at a different electrical potential, perhaps due to a different stress concentration. Electrons will flow from the anode (the area of higher potential), to the cathode, through the metal and then return to the anode through the water to complete the circuit.

At the anode the steel will go into solution. $Fe \Rightarrow Fe^{2+} + 2e^-$. At the cathode water may be dissociated to form hydrogen. This will form an insulating layer and the corrosion will stop. This corrosion mechanism is hydrogen evolution.

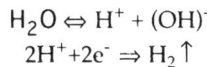

$$H_2O \Leftrightarrow H^+ + (OH)^-$$
$$2H^+ + 2e^- \Rightarrow H_2\uparrow$$

If the boiler water contains dissolved oxygen or is acidic the insulating layer of hydrogen will be removed and the corrosion will continue. The electrons will combine with the water and oxygen at the cathode to form hydroxyl irons. This corrosion mechanism is oxygen absorption. $O_2 + H_2O + 2e^-$. The metal and hydroxyl irons are free to combine to form rust as follows:

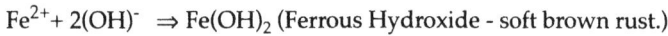

$Fe^{2+} + 2(OH)^- \Rightarrow Fe(OH)_2$ (Ferrous Hydroxide - soft brown rust.)

This may be further oxidised to form hard rust, as seen in pitting scabs:

$4Fe(OH)_2 + 2(H_2O) + O_2 \Rightarrow 4Fe(OH)_3$ (Ferric Hydroxide - hard rust.)

To prevent corrosion in a boiler the water must be both free of oxygen and maintained in an alkaline condition.

Chemicals can be applied to remove the oxygen but these tend not to be used in low pressure boilers. Removal of oxygen is more important in high pressure boilers, where the dissolved oxygen content should be less than 0.02ppm. Chemicals can also be used to maintain the water in an alkaline condition, usually sodium hydroxide in a combined chemical treatment. The boiler water should have a pH of about 11, or a P alkalinity of 150-200ppm.

Oxygen and carbon dioxide will be absorbed at the hot well. This can be reduced by keeping the hot well at a high temperature such as 85–95°C. Oxygen scavenging chemicals such as 'hydrazine' can also be used.

Sludge

When suspended matter in a boiler reaches a high level it will precipitate as sludge. On hot surfaces this may form a baked on layer and reduce the heat transfer. Boiler chemicals contain dispersants that prevent suspended matter from precipitating out and regular blowdown keeps suspended matter at a low level. Dissolved solids should be kept below 2000ppm.

Oil can also be a problem in boilers. The usual source is from leaking tank heating coils, which may be detected with sensors or be visible at the hot well or observation tank, if fitted. The source of the leak should be located and isolated as the oil will form carbonaceous deposits on heating surfaces.

Oil must be removed by boiling out. Caustic soda can be used, if mixed with fresh water in the ratio 1:500, or chemicals are available from the normal chemical treatment supplier. The boiler should be filled to its normal level and then slowly fired to raise the pressure to about 75 per cent of the normal pressure. This should be maintained for about 24 hours and then the boiler should be drained and flushed out with fresh water. If any oil deposits remain the process should be repeated.

4.5   *Thermal Fluid Systems*

Thermal fluid systems have been used on board ship for the past thirty years, however, despite their many advantages they have yet to overtake steam as the main source of heating.

The systems are relatively simple and use an oil fired heater to raise the temperature of the fluid which is pumped around the system by a single stage centrifugal pump. An expansion tank is fitted at the highest point in the system to take up expansion of the fluid. This tank also prevents air and vapour locks forming. A tube stack is fitted in the exhaust uptake to heat the fluid when the main engine is running.

The fluid is circulated at a low pressure as pressure is only required to circulate the fluid around the vessel and prevent contamination in the event of a coil leak. The relatively low operating pressure means that cheaper and more simplistic pipework can be used than with conventional steam systems. Typical circulation temperatures are between 150–250°C.

Thermal fluids are usually paraffinic or napthenic mineral oils with a low viscosity and high thermal conductivity and stability. Synthetic fluids are also used. The fluid remains in a liquid state and does not cause corrosion or scale so requires no chemical treatment. Operating costs are lower as continual blowdown and make-up are not required, however, if a coil leaks and fluid is lost the cost of refilling the system is expensive.

Maintenance of the thermal fluid system is fairly simple. Valve and pipework maintenance is virtually negligible and most of the routine maintenance involves the circulating pumps and boiler burner.

Thermal fluid samples should be taken at three month intervals and sent for analysis to check the flashpoint. This should normally be well above 200°C but if the fluid overheated at any time, perhaps due to a fault in the boiler thermostat or low circulation, the fluid can 'crack' and form lighter, lower flashpoint components. The flashpoint of the whole charge will then be lowered, and could reach a dangerous level.

5 Electrical Machinery

5.1 *General*

Requirements for the design and installation of electrical equipment are covered in the SOLAS 1974 and 1981 amendments. Classification Societies may also have their own additional requirements.

Regulations require the main generators to be capable of supplying all services to maintain a vessel in normal operation, without the need to use an emergency power source. The minimum number of generators must be such that if one goes out of use, the remaining generator(s) will be capable of providing the normal power requirements.

Electrical equipment must be maintained in good condition as it can be a source of fire if neglected. Almost 16 per cent of shipboard fires are caused by electrical faults, although, many ships no longer carry a dedicated electrician and maintenance often gets carried out by the ship's engineers. With reduced crewing levels and shorter periods of time spent in port, electrical maintenance may not be given the priority it needs.

Maintenance is about preventing faults from developing and causing damage. This involves:

-   Cleaning the equipment. Blowing out dust from motors and starters, cleaning air intake grilles, degreasing motor windings, etc.
-   Regularly checking and recording the insulation readings.
-   Checking the tightness of connections and ensuring enclosures are tight to dirt, dust and water, particularly on deck.
-   Monitoring the bearings.
-   Inspecting the contacts, switchgear and brushgear.
-   Checking the settings of protective devices.

Distribution

Shipboard 440V electrical distribution systems are normally fully insulated and totally insulated from the ship's hull as the neutral point is not earthed. The reason for this is that when an earth fault develops machinery will continue to run. Metal parts, such as enclosures and casings, are still earthed to the ship's hull so that personnel do not get a shock. Instead of blowing a fuse or circuit breaker, the fault is detected by earth fault indicators.

Safety

Electricity kills. Before commencing work on any electrical machinery ensure that it is isolated. Do this yourself - you are going to work on the machine so don't rely on anyone else. Remove the fuses and keep them with you. If possible lock the isolator in the open position and post a notice advising other personnel that the equipment is being worked on. Always check the equipment is 'dead' with a meter. A 'permit to work' is required in order to carry out work on electrical equipment above 1000V. In case of electric shock:

- Do not touch the victim.
- Isolate the electrical supply, if this can be done rapidly. If not then try to remove the victim by pulling them clear with a jacket or a length of wood or rope - do not touch the victim directly.
- Once the victim is clear, check they are breathing and carry out artificial respiration if necessary.
- Seek medical attention.

5.2   AC Motors and Generators

AC Motors

The three phase induction motor consists of a stator, which houses the three phase windings in slots, and a laminated steel rotor. Wound rotors are now rarely used on board ship.

AC Generators

The three phase ac generator is usually a salient pole machine consisting of a laminated stator with a three phase winding carried in slots and a rotor of 8 or more poles energised by field coils, whose connections are brought out to slip rings on the shaft.

The rotor field current is supplied via an exciter. This can be a dc generator mounted on the end of the shaft (rotary exciter); supplied from the three phase supply via a rectifier (static exciter) or, in the case of a brushless machine, via an ac exciter and rectifier.

Machinery such as compressors or cargo gear is started and stopped often. On an unregulated generator a sudden load increase would result in a voltage dip of approximately 30 per cent. To reduce this to an acceptable level an automatic voltage regulator (AVR) is fitted. This limits the transient voltage dip to 15 per cent and maintains the steady state voltage to within ±2.5 per cent.

Parallel Operation of AC Generators

Before connecting another generator to the busbars three criteria must be met:
1. The voltage of the incoming machine must be the same as the busbar voltage.
2. The frequency of the incoming machine must also be the same as the busbar frequency.

3. The voltage of the incoming machine must be in phase with the busbar voltage.

Guidelines for parallel operation are as follows:

1. Set the generator selector on the switchboard to the incoming machine. This will switch on the synchroscope which will be rotating in either a clockwise or anti-clockwise direction.
2. The voltage should be automatically adjusted by the AVR.
3. The frequency of the incoming machine will need to be adjusted, usually upwards. This is done by adjusting the speed regulator to increase the engine speed. If the synchroscope is rotating anti-clockwise increase the speed regulator and vice versa.
4. The synchroscope pointer will slow down and the speed regulator should be adjusted until the pointer is rotating slowly in the clockwise direction.
5. When the pointer is at the 5 to 12 position the circuit breaker should be closed. The voltage of the incoming machine and the busbars are in phase at the 12 o'clock position. If the incoming machine is not in phase there may be an interlock that prevents the breaker closing.
6. The incoming machine can share some of the load by increasing the speed regulator of the incoming machine while decreasing the speed regulator of the machine(s) already connected.

On many modern vessels the above operations are carried out automatically.

5.2.1   Faults and Effects

Generators

Prime Mover Failure

If only one generator is running the vessel will suffer a blackout. If running in parallel, the failed generator will try to run as a synchronous motor to drive the engine, causing the reverse power trip to operate. The remaining generator will then take full load. This may cause overload again, resulting in a blackout.

Excitation Failure

If a single generator is running there will be a loss of voltage causing a blackout. If the generators are running in parallel they will supply a large leading current, causing the overcurrent trip to operate.

On a brushless generator the rectifier diodes on the shaft can also fail. If one fails in an open circuit the field current will reduce. The AVR will try to compensate for this reduction causing a risk of the field winding overheating. If a diode short circuits there will be an increase in excitation current which could also cause damage to the field winding.

Motors

Single Phasing

This is when an open circuit occurs on one line of a three phase winding causing the motor to act as a single phase machine. This can be due to a blown fuse, a broken wire on one line, a loose terminal, a dirty contactor, or a burnt out winding.

Effects on a stopped motor:

- The motor will not start.
- The motor takes a large current.
- There is a loud humming noise.
- Overheating takes place rapidly.

Effects on a running motor:

- The motor continues to run with an increased current (2.5 x I_{FL}).
- The motor eventually overheats and can burn out.

Overheating

This may be due to the following factors:

- Overload.
- Dirty windings or dirty air intake grilles.
- Damaged cooling fan.
- High ambient air temperature or inadequate ventilation.
- Incorrect voltage or frequency.

Maintenance

Maintenance is about preventing faults from developing and causing further damage. It means ensuring the following steps are taken:

- Machinery and equipment are clean.
- Enclosures are tight to prevent water and dirt ingress.
- Checking insulation readings.
- Checking bearings.
- Ensuring all fixings are tight.
- Inspecting contactors and switches for pitting, excessive burning or arcing.
- Checking the settings of protective devices.

5.3 Batteries

Batteries for emergency use should be situated above the bulkhead deck outside the machinery space and aft of the collision bulkhead. Regulations cover where the batteries can be placed, the charging arrangements, ventilation requirements, etc.

Batteries can be either lead acid or alkaline but both types require the same amount of care and attention.

Alkaline Batteries

Alkaline batteries suffer no change in the electrolyte specific gravity between full charge and no charge, therefore, the state of charge must be determined by measuring each cell voltage. The specific gravity of the electrolyte is about 1.17 but this will deteriorate over a long period of time and should be replaced when it falls to 1.145.

Always measure the voltage when the battery is on load as the no load voltage is higher and can give a false indication that the battery is satisfactory.

Lead Acid Batteries

A lead acid battery will suffer a change in specific gravity of the electrolyte from 1.28, when fully charged, to 1.11 when discharged. The specific gravity is affected by temperature so this must be corrected to 25°C (±0.007 per ^{o}C). Temperatures above this will reduce the battery capacity and increase the rate of evaporation of the electrolyte. Lead acid batteries must be left on a trickle charge.

Both types of battery give off hydrogen when charging so there may be a potentially explosive atmosphere in the battery locker.

Maintenance

It is easy for batteries to get overlooked as they are tucked away in a locker on deck. Maintenance is simple and should be a weekly routine, along with all the other emergency equipment. Ensure the battery room is well ventilated and that vent flaps are free. Filters should be fitted to the vent openings if the atmosphere is dusty. Measure and record cell voltage and specific gravity.

Check electrolyte levels and remember that any loss will increase in warmer climates. Always use rubber gloves, an apron and goggles when filling batteries with new electrolyte as it can burn - particularly alkaline battery electrolyte. Ensure the tops of the batteries are clean and dry and that all terminals are tight.

5.4   Protective Devices

Ships electrical systems are protected by various devices, such as fuses and circuit breakers, to prevent damage to the components and insulation and also to minimise the risk of fire. The connection between the generators and the switchboard, and the steering gear circuits are the two exceptions. The latter are protected against short circuit by an overload alarm.

Overcurrent Protection

This is to prevent damage to an electrical insulation from a sustained overcurrent. A time delay is built in to allow transient overloads to pass without tripping the breaker, such as during startup. Some motors may also have a differential overload relay which will protect against single phasing.

Short Circuit Protection

A short circuit will generate a very large fault current which will damage generators and the distribution system, if not interrupted quickly. The short circuit protection device is similar to the overcurrent relay but has no built in time delay and is set at a high value.

No-Volt Protection

A no-volt protection device prevents a circuit breaker from being closed, so that a stopped generator cannot be connected onto live busbars.

Reverse power protection

If two or more generators are connected to the busbars and the engine of one of them fails, the failed generator will no longer supply power and will try to run as a synchronous motor. This means it would take power from the switchboard, which will probably cause a blackout and could also damage the engine. A reverse power relay will open the generator circuit breaker alleviating this problem.

Regulations state that if two or more generators are required to operate in parallel, a reverse power protection device must be installed.

5.5 Loss of Power

If a blackout occurs for any reason the ship's staff should be familiar with the procedure for re-establishing electrical power. The main cause of a blackout is usually overloading, due to starting another machine when there is insufficient spare capacity to accommodate it. It can also be caused by generator engine failure.

A common cause of engine failure is dirty fuel filters particularly in bad weather when sediment in fuel tanks becomes stirred up. An indication of this is a fluctuation in busbar frequency and, if this occurs, the engine should be checked to ensure exhaust temperatures and rpm are steady.

Once the main power has been lost the emergency generator should startup and supply emergency lighting. If the diesel engine has failed then a standby generator will have to be started. The reason for the engine failure should then be determined.

If the generator has tripped on overload and the engine is still running, the power should be restored as quickly as possible as the engine may lose its cooling supply.

The circuit breaker will most probably need to be closed manually and the synchronising system may have to be switched to bypass. Usually there is a locked automatic synchronising switch on the front of the switchboard.

Once the power has been restored all the main pumps for the engine should be started and the engine temperatures stabilised for a period, before starting the main engine. *Note: Some pumps may have sequential starting and will start up automatically once power is restored.*

5.6   Earth Faults

As previously stated, most ships have an insulated distribution system. If an earth fault develops on this system there is no short circuit to earth that could cause a fuse to blow, as would occur in an earthed system. The system should continue to function normally. If another earth fault develops the two earths create a short circuit which causes protective devices to operate. This results in an interruption to services, some of which may be essential, so it is important that earth faults are quickly found and removed. An insulated system therefore requires a fault on two different circuits before an earth current will operate protective devices (an earth fault is said to occur when the insulation resistance falls below 1 megohm).

The most common causes of earth faults are damp, dirt, aging or damaged insulation, and high temperature.

Damp

This is the most common cause of earth faults, particularly on deck machinery or lighting circuits. The cause is usually damaged, missing or improperly fitted cable glands or terminal box seals allowing water to enter. Sea water can cause problems as salt can build up leaving deposits on contacts or terminals. This buildup must be removed, even if it means flushing with fresh water first. Machinery in deck houses usually have space heaters to prevent condensation and these should be switched on.

Dirt

As with damp, dirt and dust can be a problem and the cause of ingress is usually the same. Motors on deck are usually totally enclosed but if they are not, particular attention needs to be paid to the cleanliness of the windings. If machinery is in an enclosed room there may be a ventilation fan and the intake should have a filter fitted.

Earth Indicators.

Regulations require that insulated systems should have a means of detecting any earth faults. This may be with an earth leakage monitor or earth lamps may also be used, although they will not detect a fault below 4000ohms.

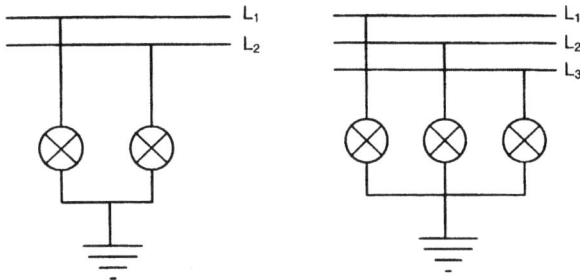

Figure 117. Earth Lamps

No faults   All lamps glow with equal brilliance.

Earth fault   The lamp on the line that has an earth will dim, or go out, depending on the severity of the fault.

Clearing Earth Faults

If an earth fault is detected it should be cleared as soon as possible. This involves isolating each circuit one at a time until the fault clears. Circuits that are more susceptible should be tested first, such as deck machinery, deck lighting circuits, the galley and all areas where water ingress is possible. The fault may be difficult to detect if several machines are running and machines may have to be changed over.

Once the faulty circuit has been detected then insulation readings should be taken, starting from the main breaker for the faulty circuit and progressing down the circuit through junction boxes, starters, terminal boxes, etc. until the fault is found. On 440V systems the insulation tester should operate at 500V. No electronics, including AVRs, should be on the circuit being tested as these will be damaged.

5.7   *Hazardous Area*

A hazardous area is defined as 'an area in which explosive gas/air mixtures are, or may be expected to be, present in quantities such as to require special precautions for the construction and use of electrical apparatus'.

Hazardous areas are divided into zones, depending upon the risk of an explosion due to a gas/air mixture.

Zone 0   An explosive gas/air mixture is continuously present, or present for long periods. An example would be oil cargo tanks.

Zone 1   An explosive gas/air mixture is likely to occur under normal operating conditions. Examples are battery rooms, paint rooms and enclosed deck spaces on tankers.

Zone 2   An explosive gas/air mixture is not likely to occur. If it does, it will only be for a short period. An example would be the open spaces on the deck of a tanker.

Depending on the zone, special electrical equipment must be used to avoid an explosion. Explosion protected equipment can be identified by the letters 'ex' followed by another letter to indicate the type of protection.

exd   Flameproof enclosure. This is of heavy construction with wide flanges to dissipate the energy of any internal explosion.

exi   Intrinsically safe. Any spark or thermal effect is not of sufficient energy to ignite an explosive mixture.

exp   Enclosure is pressurised with an inert gas.

exe   Increased safety. The design is such that any sparking of contacts, brushes, etc. will be eliminated.

exo   Equipment is immersed in an oil filled tank.

exq   As above, but with sand instead of oil.

5.7.1   Maintenance of Electrical Equipment

Electrical equipment and machinery needs to be maintained irrespective of its location.

When working in hazardous areas all equipment must be electrically dead and earthed to ensure no currents occur from any capacitative effect. If possible the area should be gas freed. All test equipment should be intrinsically safe and tools should be suitable for use in a hazardous area, that is they must not create a spark if dropped.

When inspecting equipment ensure all flame paths, such as terminal box flanges, are free from damage and any signs of pitting or corrosion. Check all bolts are present and that they are correctly tightened down.

5.8   Shore Supply

The shore power connection box is usually located in a locker on deck, often where the emergency generator is situated.

The three shore supply cables should be brought in through an opening or hatch, not over a door sill where they are at risk of damage. The cables should be in a good condition and have no damaged insulation. The cables should be connected up to the ship's shore power connection box and the dockyard can then switch on the supply. The ship's shore supply breaker should remain open.

The supply should then be checked for correct phase rotation, voltage and frequency. If the phase rotation is incorrect the supply should be turned off and two of the cables changed over.

The ship's main generator circuit breaker is then opened (tripped) and the shore supply circuit breaker closed. *Note: There will be a blackout during the changeover and the ship's staff should be notified in advance.*

The shore supply power will probably be much less than the normal ship supply and care should be taken to monitor the load when starting machinery to prevent the shore supply circuit breaker from tripping.

When disconnecting the shore supply, the shore supply breaker should be opened (tripped) first, which will cause a blackout again. Once the shore supply breaker has been tripped the main generator can be started and the main circuit breaker closed. Again, inform all ship's staff when this will occur as well as any dock workers who may be working on board ship.

Note: If the vessel is in drydock, this operation is carried out while flooding the dock, when the water level is above the cooling water pump sea suctions.

6 Waste Treatment & Disposal

6.1 Waste Generation

Solid and liquid wastes are generated on board ship from a variety of sources:

- Food waste from crew and passengers.
- Toilets.
- Packaging (boxes, plastic bags, etc.).
- Bilges.
- Sludge and oily waste from tank drains, purifiers, back flushing filters, etc.
- Tank cleaning.

Some of this waste may be biodegradable such as food waste but other waste, such as oily sludge and plastic, is persistent. The result is that some waste can be disposed of at sea while some must be disposed of ashore or on board and recent years have seen a growing awareness of the damage done to the marine environment by uncontrolled dumping of shipboard waste.

Marine law regarding pollution is covered in the six annexes to IMO's International Convention for the Prevention of Pollution from Ships 1973, as modified by the protocol of 1978, often referred to as MARPOL 73/78. A copy of this should be kept on board and ship's staff should be familiar with the requirements of the following annexes.

6.2 MARPOL Regulations

Annex I

This covers pollution by oil. The oil content of machinery space discharges for tankers and all other vessels over 400grt must not exceed 15ppm. The vessel must also be en route, outside any special area and equipped with an oil discharge monitoring and control system. For the purposes of Annex I special areas are the Red Sea, Mediterranean Sea, Baltic Sea, Black Sea, Gulfs Area, Gulf of Aden and Antarctica. Northwest European waters have been designated 'special areas' effective as of August 1st, 1999. These area include the English Channel, North Sea, Irish Sea and part of the Northwest Atlantic Ocean.

Processed bilge water can be discharged in special areas provided the vessel is en route, the bilge water does not originate from a cargo pump room bilge or been mixed with cargo oil residues, the oil content of the discharge does not exceed 15ppm, and on the provision that the discharge will be automatically stopped if the oil content exceeds 15ppm.

The following sign must be placed in prominent positions such as on the oily water separator unit and at the bunker stations.

DISCHARGE OF OIL PROHIBITED

THE FEDERAL WATER POLLUTION CONTROL ACT PROHIBITS THE DISCHARGE OF OIL OR OILY WASTE INTO OR UPON THE NAVIGABLE WATERS OF THE UNITED STATES, OR THE WATERS OF THE CONTIGUOUS ZONE, OR WHICH MAY AFFECT THE NATURAL RESOURCES BELONGING TO IT.

APPERTAINING TO, OR UNDER THE EXCLUSIVE MANAGEMENT AUTHORITY OF THE UNITED STATES, IF SUCH DISCHARGE CAUSES A FILM OR DISCOLOURATION OF THE SURFACE OF THE WATER, OR CAUSES SLUDGE OR EMULSION BENEATH THE SURFACE OF THE WATER, VIOLATORS ARE SUBJECT TO SUBSTANTIAL CIVIL PENALTIES AND/OR CRIMINAL SANCTIONS INCLUDING FINES AND IMPRISIONMENT.

THE USE OF CHEMICAL DISPERSANTS TO DEAL WITH OIL DISCHARGES IS PROHIBITED IN THE UNITED STATES NAVIGABLE WATERS WITHOUT THE PRIOR CONSENT OF THE USCG ON-SCENE CO-ORDINATOR.

Annex II

This covers pollution by liquid noxious substances carried in bulk. Discharge of residues is only allowed at shoreside reception facilities.

Annex III

This covers pollution by harmful substances, in packaged form or in containers. The general requirements involve the issuing of detailed standards on packing, marking, labelling, documentation, stowage, quantity limitations, etc.

Annex IV

This applies to all ships over 200grt and limits the discharge of sewage into the sea. Disinfected and comminuted sewage may not be discharged closer than four nautical miles from land. Other forms of sewage may not be discharged closer than 12 nautical miles.

Guidelines for effluent discharge levels are published by the Marine Environmental Protection Committee of IMO. They are as follows:

- 250 faecal colliforms / 100ml (max).
- Total suspended solids 50mg / litre.
- BOD5 50mg / litre.

BOD5 is the five day biological oxygen demand and is a measure of oxygen consumption. It should be noted that the colliform count is lower (200) in the Baltic Sea, Black Sea, Great Lakes and the St. Lawrence Seaway.

Annex V

This annex covers the disposal of garbage and applies to all ships. It prohibits the dumping of any form of plastic, including synthetic ropes and plastic garbage bags. Dunnage and packing material that will float may not be dumped less than 25 nautical miles from land. Food waste and other garbage such as paper, rags, glass, metal and similar refuse cannot be discharged within 12 nautical miles of land unless it has been ground up, then the minimum distance is 3 nautical miles.

For special areas the disposal of any garbage other than food waste is prohibited and even then it must be more than 12 nautical miles from land. For the purposes of Annex V, special areas refer to the Mediterranean Sea, Baltic Sea, Black Sea, Gulfs Area, North Sea, Gulf of Mexico, the Caribbean and Antarctica. Table 16 summarises the vessel requirements.

GARBAGE TYPE	OUTSIDE SPECIAL AREAS	IN SPECIAL AREAS
Plastics - Including synthetic ropes, fishing nets and plastic bags	Disposal Prohibited	Disposal Prohibited
Floating dunnage, lining and packing material	> 25 miles offshore	Disposal Prohibited
Paper, rags, glass, metal, bottles, crockery and similar refuse	> 12 miles offshore	Disposal Prohibited
All other garbage including paper, rags, glass etc. not comminuted or ground	> 3 miles offshore	Disposal Prohibited
Food waste not comminuted or ground	>12 miles offshore	> 12 miles offshore
Food waste comminuted or ground	> 3 miles offshore	> 12 miles offshore
Mixed refuse types	*	*

Table 16. Summary of Marpol Annex V

* When garbage is mixed with other harmful substances which have different disposal or discharge requirements, the more stringent disposal requirements shall apply. *Note: Ground or comminuted waste must be small enough to pass through a 25mm mesh filter.*

Annex VI

This annex is concerned with controls to limit air pollution from ships. At the time of writing it has yet to be ratified, although it is expected to come into force on 1st January 2000. Its main aim is to reduce oxides of sulphur (SO_x) emissions, which are a main cause of acid rain. The annex will set a cap of 4.5 per cent on the sulphur content of marine fuels. There will also be special areas where the cap will be 1.5 per cent. At present this includes the Baltic Sea, however, the North Sea may also be included.

The SO_X regulations are rather controversial. On the one hand there are complaints that it doesn't go far enough because most marine residual fuels have sulphur contents in the region of three per cent. On the other side of the argument, against a cap, is the fact that a vessel will have to be able to store and treat two different residual fuel grades and low sulphur fuel (<1.5 per cent) is not widely available. In addition there is the problem of how it will be policed.

The annex also covers oxides of nitrogen (NO_X) emissions. Diesel engines installed in new vessels will have to have a certificate to show they comply with the NO_X emissions. Oxides of nitrogen are the so called greenhouse gases that deplete the ozone layer. In addition they cause smog and acid rain. Diesel engines produce oxides of nitrogen during combustion and most of the nitrogen comes from the combustion air.

NO_X formation is a function of temperature and the time held at that temperature. As a result, slow speed diesel engines tend to produce more NO_X due to the longer residence time of the gases in the combustion chamber.

Dealing with NO_X emissions is fairly complex and there are two principal methods for NO_X reduction.

- Primary - NO_X is reduced during combustion.

- Secondary - NO_X is removed from exhaust gases.

It is anticipated that primary methods will be sufficient to meet expected NO_X regulations. Engine manufacturers are carrying out much research into the most efficient methods of reducing NO_X emissions and a few of these methods follow. [1]

Primary Methods

These work on the basis that reducing the flame temperature reduces the NO_X formation.

- Retard fuel injection.
    - This will reduce P_{max}, which effects combustion temperature. Efficiency is also reduced, however, which increases the specific fuel oil consumption(spoc).
- Increase compression ratio.
    - This can only be achieved by a physical change to the combustion chamber geometry, however, an effective increase can be achieved by adjusting the valve timing or increasing scavenge pressure. The aim is to achieve no pressure rise after combustion, i.e. constant pressure combustion. The best approach is to retard fuel injection and increase compression pressure.

---

1. For Further information please refer to IMarE's Marine Engineering Practice Series, Volume 3, Part 20, *Exhaust Emissions from Combustion Machinery* by A A Wright.

- Reduce excess air ratio.
  - Reducing the amount of oxygen will reduce the amount of NO_X formed. Unfortunately, reducing the amount of air increases the thermal load on the engine.
- Reduce scavenge air temperature.
  - This has the effect of reducing the combustion temperature but is limited by the temperature of the cooling water.
- Fuel distribution.
  - This is influenced by the number of injectors, spray pattern, size of fuel droplets and the swirl, which all influence heat release.
- Water injection.
  - Water has a cooling effect as it absorbs heat when evaporating.
  - Adding to scavenge air. This may affect the cylinder oil film on the liner.
  - Fuel/water emulsion. Larger fuel pumps and injectors are required for full load operation as the volume of water has be added to volume of fuel. The engine must therefore run on emulsion at all times.
  - Direct injection. Wärtsilä NSD consider this the most favourable method, giving NO_X reductions of up to 60 per cent.
- Exhaust gas recirculation.
  - Reduces the oxygen content of combustion gases.

Secondary Methods

The most preferred method is selective catalytic reduction (SCR). A catalyst such as ammonia is used to reduce NO_X, however, the use of residual fuels, with their relatively high sulphur contents, can cause problems for catalytic converters. Research is still continuing into the use of catalytic converters on board ship.

Air pollution from incinerators is included in the Annex as well as the types of material that can be burnt in them. PVC cannot be burnt in incinerators which do not have an IMO Certificate. On ships with older incinerators without IMO Certificates, all PVC must be disposed of ashore. From the year 2000 incinerators fitted to all new ships will need to have an IMO Certificate.

The Annex also covers ozone depleting substances and is in line with the Montreal Protocol.

All Annexes, except IV and VI, are in force. The remaining Annexes were amended in 1996 so that port state control inspectors could ensure a ship's staff are competent in operating the waste disposal equipment.

6.3   Waste Disposal

Marpol regulations require that vessels have a garbage disposal plan. A garbage record book must be kept, in a similar manner to the oil record book. Procedures must be set in place for the collection, sorting, processing and disposal of garbage and there must be a designated person to implement the plan. This will ensure that all staff are fully aware of how, when and where waste can be disposed of. Disposal of garbage, either overboard, ashore or in the incinerator, must be logged in a garbage record book along with the date, time and ship's position.

This regulation came into effect from July 1997 for newbuildings over 400grt. Existing vessels should have complied by July 1998.

Sewage and bilge water are disposed of via the sewage treatment unit and the oily water separator system respectively.

Sludge and Oil Waste

Sludge produced by the purifiers must be either incinerated or disposed of at a shoreside reception facility. Disposal of any waste oil must be recorded in the oil record book - as it could be inspected by Port State Inspectors. If waste oil is disposed of ashore ensure a receipt is obtained from the shore reception facility and keep it in the oil record book.

Sludge should not be disposed of by pumping it into fuel tanks as it will be removed again by the purifiers. If incinerating waste oil it should be pumped to the incinerator mixing or settling tank and then heated prior to incinerating. During this time any water can be drained from the oil.

Solid Waste

There are different disposal requirements for different types of garbage. Waste should be segregated into separate bins, for plastics, paper, rags, tins, glass, etc., and food waste. All plastics should be either incinerated or disposed of ashore. In special areas only food waste can be disposed of at sea, and only when more than 12 miles from the shore. The alternative is for all waste to be incinerated. The high temperature within the incinerator produces a sterile ash which can then safely be disposed of.

Shredders and compactors help reduce the volume of waste prior to disposal. Shredding waste can reduce the volume by up to 90 per cent prior to incinerating. Compacting is used prior to disposal ashore where the reduced volume reduces disposal costs.

6.4 Oily Water Separator Systems

Bilge water discharges must contain less than 15ppm of oil under the current IMO requirements. To achieve this the bilge water must be passed through a separating system. This is usually a two or three stage separation process. Primary separation is achieved by a mixture of filtration and centrifugal separation, followed by a coalescer/filter unit.

The overboard discharge is sampled through an oil content meter. If the oil content exceeds 15ppm the overboard discharge valve closes and the discharge is returned to the bilge tank. Problems have been experienced in achieving 15ppm discharge and these problems are usually caused by detergent cleaners becoming mixed with the bilge water. This holds the oil suspended in a fine emulsion which cannot be separated out. A typical separator system is shown in Figure 118.

Prior to startup and shutdown the oily water separator must be flushed through with clean sea water. Sludge tanks should never be pumped through the system as the sludge will coat the insides and rapidly clog up filters and coalescers.

PROBLEM	CAUSE	ACTION
Water in oil discharge	Oil or sludge on oil sensor	Remove and clean sensor
	Sensitivity of oil sensor needs adjusting	Adjust
	Inadequate separation due to: Sludge inside unit	Open up and clean
	Too much oil in bilge water	Apply steam heating
	Bilge water contaminated with detergent	Flush through with sea water
High oil content in water discharge (1st Stage)	Sensitivity of oil sensor needs adjusting	Adjust
	Oil discharge valve not fully open	Check valve operation
High oil content in water discharge (2nd/3rd Stage)	Dirty Filters/coaleser	Check condition and clean or replace
	Bilge water contaminated with detergent	Flush through with sea water
	Problem with oil content meter	Check condition of discharge water. If clear, flush through meter cell with clean water and clean cell.
High pressure in separator	Dirty filters or coalescer	Check condition and clean or replace
	Outlet valve closed	Check valve setting

Table 17. Oily Water Separator Problems

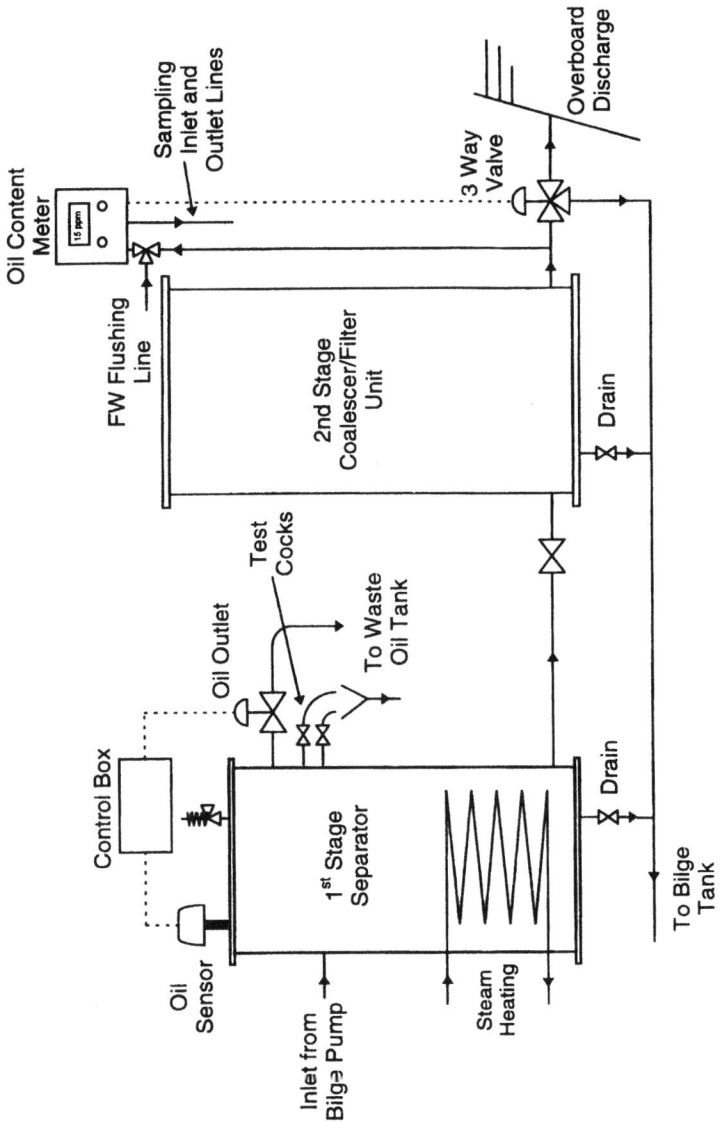

Figure 118. Typical Oily Water Separator System

6.5 Sewage Plant

Sewage plant can be divided into four types.

1. Chlorination.
2. Physical/Chemical Treatment.
3. Electrochemical.
4. Biological.

Chlorination

This is a relatively simple system. The sewage is kept in a storage or holding tank where it is disinfected before being pumped out at sea or to a shore facility.

Physical/Chemical Treatment

Solids and liquids are separated through a coarse screen filter. The liquid is further clarified by being filtered through fine filters. It is then disinfected before it is discharged overboard. Solids are retained for dumping out at sea or to a shore facility.

Electrochemical

The advantage of this system is that it is compact and lighter than other systems. It can also be quickly started and stopped as required.

The sewage is passed though an electrocatalytic cell which oxidises the sewage. The salt in the sea water that is used for flushing the toilets gets converted to sodium hypochlorite, a disinfectant, by the cell. No residual sludge is left after the process.

Biological

Raw sewage is passed to an aeration chamber and air is supplied through diffusers. This promotes the action of aerobic bacteria which break down the sewage into carbon dioxide, water and inorganic waste. Figure 119 shows a typical biological treatment plant.

Sewage passes into the clarification chamber. Any solids that settle out are returned via an air lift to the aeration chamber which ensures they are fully broken down. The same applies to any surface scum. A small vane type air compressor supplies the air for the diffusers and air lifts.

The clear liquid then passes through the chlorinator where the liquid is disinfected, into the chlorination chamber. This chamber has float switches, which control the discharge pump, and a high level alarm.

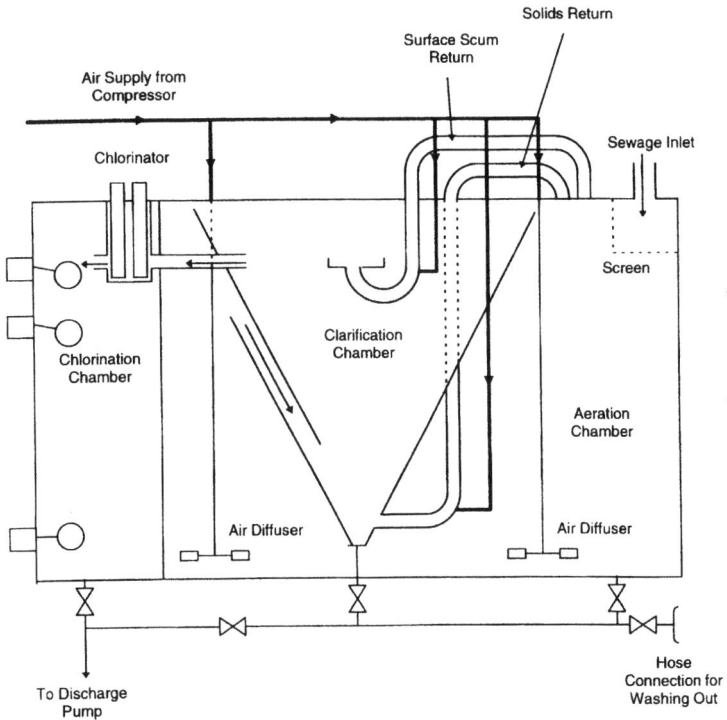

Figure 119. Biological Sewage Treatment Unit

Although the sewage treatment unit runs automatically, without regular maintenance the unit will not function correctly and anaerobic bacteria may promote the formation of hydrogen sulphide and methane, both of which are hazardous.

The chambers need to be cleaned out occasionally to remove any accumulated matter. The aeration diffusers should be checked to ensure they are clear and that air is bubbling from them. The air lift returns should also be checked to make sure they function correctly. These usually have a clear plastic pipe so that the sludge can be seen returning to the aeration chamber.

The internal tank coating should be inspected for any signs of cracking or blistering.

When cleaning out a sewage unit rubber gloves and a mask should be worn. After overhaul the external surfaces of the unit and the surrounding area should be washed down with disinfectant. Hands should also be thoroughly scrubbed and overalls washed.

List of Abbreviations

AVR	automatic voltage regulator
bhp	brake horse power
bmep	brake mean effective pressure
CCAI	calculated carbon aromaticity index
CFC	chlorofuorocarbon
COP	coefficient of performance
CPP	controllable pitch propeller
cSt	centistoke
dwt	deadweight tonnes
EC	European Community
grt	gross registered tonnage
GWP	global warming potential
HCFC	hydrochlorofluorocarbon
HFC	hydrofluorocarbon
HP	high pressure
IFO	intermediate fuel oil
IMO	International Maritime Organisation
ISO	International Standards Organisation
LO	lubricating oil
LP	low pressure
mcr	maximum continuous rating
mep	mean effective pressure
NO_X	oxides of nitrogen
SCR	selective catalytic reduction
sfoc	specific fuel oil consumption
SO_X	oxides of sulphur
TEWI	total equivalent warming impact
UMS	unmanned machinery space
VEC	variable exhaust closing
VIT	variable injection timing
VLCC	very large crude carrier

Table of Conversions

LENGTH

cm	m	km	inch	feet	yard	mile	n. mile
1	1×10^{-2}	1×10^{-5}	0.3937	3.281×10^{-2}	1.094×10^{-2}		5.4×10^{-4}
100	1	1×10^{-3}	39.37	3.281	1.094	6.214×10^{-4}	0.54
1×10^{5}	1000	1	39.37×10^{3}	3281	1094	0.6214	0.54
2.54	2.54×10^{-2}	2.54×10^{-5}	1	0.0833	0.02778		
30.48	0.3048	3.048×10^{-4}	12	1	0.333	1.894×10^{-4}	1.646×10^{-4}
91.44	0.9144	9.144×10^{-4}	36	3	1	5.682×10^{-4}	4.937×10^{-4}
	1609	1.609		5280	1760	1	0.8690
	1852	1.852		6076	2025	1.151	1

VOLUME

litre	m³	barrels	Imp gallon	US gallon
1	1×10^{-3}	6.29×10^{-3}	0.22	0.264
1000	1	6.29	220	264
159.0	0.1590	1	34.97	42
4.546	4.546×10^{-3}	0.02859	1	1.201
3.785	3.785×10^{-3}	0.02381	0.8327	1

POWER

kW	PS	HP
1	1.3596	1.34
0.7355	1	0.9863
0.7457	1.0139	1

PRESSURE

Pascal (Pa)	bar	kg/cm²	lb/in²	atm	mm Hg	in Hg	m H₂O
1	1×10^{-5}	1.0197×10^{-5}	1.45×10^{-4}	9.869×10^{-6}	7.501×10^{-3}	2.953×10^{-4}	1.012×10^{-4}
1×10^{5}	1	1.0197	14.50	0.9869	750.1	29.53	10.12
9.807×10^{4}	0.9807	1	14.22	0.9678	735.6	28.96	10
6895	0.06895	0.07031	1	0.06805	51.71	2.036	0.7031
1.013×10^{5}	1.013	1.033	14.70	1	760	29.92	10.34
133.3	1.333×10^{-3}	1.360×10^{-3}	0.01934	1.316×10^{-3}	1	0.03937	0.0136
3386	0.03386	0.03453	0.4912	0.03342	25.40	1	0.3453
9806	0.09806	0.1	1.422	0.09678	73.55	2.896	1

Note: 1 Pascal = 1 N/m²

Index

Index

Notes

Notes

Notes

Notes

Notes